Muriel Seltman
Robert Goulding

Thomas Harriot's Artis Analyticae Praxis

An English Translation with Commentary

Springer

Sources and Studies in the
History of Mathematics and Physical Sciences

K. Andersen
Brook Taylor's Work on Linear Perspective

K. Andersen
The Geometry of An Art

H.J.M. Bos
Redefining Geometrical Exactness: Descartes' Transformation of the Early Modern Concept of Construction

J. Cannon/S. Dostrovsky
The Evolution of Dynamics: Vibration Theory From 1687 to 1742

B. Chandler/W. Magnus
The History of Combinatorial Group Theory

A.I. Dale
History of Inverse Probability: From Thomas Bayes to Karl Pearson, Second Edition

A.I. Dale
Pierre-Simon Laplace, Philosophical Essay on Probabilities, Translated from the fifth French edition of 1825, with Notes by the Translator

A.I. Dale
Most Honourable Remembrance: The Life and Work of Thomas Bayes

P.J. Federico
Descartes On Polyhedra: A Study of the *De Solidorum Elementis*

B.R. Goldstein
The Astronomy of Levi Ben Gerson (1288–1344)

H.H. Goldstine
A History of Numerical Analysis from the 16th Through the 19th Century

H.H. Goldstine
A History of the Calculus of Variations From the 17th Through the 19th Century

G. Graßhoff
The History of Ptolemy's Star Catalogue

A.W. Grootendorst
Jan De Witt's Elementa Curvarum Linearum, Liber Primus

T. Hawkins
Emergence of the Theory of Lie Groups: An Essay in the History of Mathematics, 1869–1926

A. Hermann/K. von Meyenn/V.F. Weisskopf (Eds.)
Wolfgang Pauli: Scientific Correspondence I: 1919–1929

Continued after Index

Sources and Studies
in the History of Mathematics
and Physical Sciences

Muriel Seltman
Greenwich University
Greenwich, London, SE9 2UG
United Kingdon
Muriel@seltman.fslife.co.uk

Robert Goulding
Program of Liberal Studies
Notre Dame University
Notre Dame, IN 46556
USA
Robert.D.Goulding.2@nd.edu

Sources and Studies Editor:
Jed Buchwald
Division of the Humanities
and Social Sciences
228-77
California Institute of Technology
Pasadena, CA 91125
USA

Mathematics Subject Classification (2000): 01AXX

ISBN 978-1-4939-0201-9   ISBN 978-0-387-49512-5 (eBook)

Printed on acid-free paper.

9 8 7 6 5 4 3 2 1

springer.com

Preface

The *Artis analyticae praxis* was published in 1631 in Latin. Until relatively recently, an English edition would not have been considered necessary since most of the people who might have been expected to be interested in the text would have been able to read it in Latin. That is no longer the case and it is fitting that an English edition should be published.

It is also the case that a considerable proportion of the readership of the present volume will not be professional mathematicians, so we have tried to produce a translation that makes the mathematical content accessible to the modern reader. This is not because the algebra is intrinsically difficult but because it is not the kind of mathematics which is a part of today's secondary school curriculum.

A further problem lies in the fact that the book in the form in which it was published in 1631 may very well not conform to Harriot's intentions for the publication of his mathematical manuscript papers. This is why the book is accompanied by a Commentary which attempts to compare it with the appropriate passages in the surviving manuscript papers.

The present work is a translation of the original text and not intended as a facsimile. The original has well over 300 errors (there may very well be more) and we have listed these at the end. In the interests of mathematical accessibility we have tried to produce a mathematically "clean" copy.

Again, for ease of reading (and printing) we have altered Harriot's sign for equality in his manuscripts to the modern version and omitted the two vertical lines between the parallels. Where it has been necessary to use the inequality signs seen in Harriot's manuscripts, we have used the modern version for the same reasons. Similarly, we have not included the ubiquitous dots appearing in the original work, which were common at the time and which separated the numerical from the literal part of an algebraic term, thus 2.x for the modern 2x.

The translation was the responsibility of Robert Goulding and Muriel Seltman was responsible for the Commentary. Both the translation and the Commentary were originally based on an M.Sc. Dissertation presented at University College, London by Muriel Seltman, but as work proceeded these influences have disappeared without trace and the present book is totally new.

We would like to acknowledge the assistance of the British Library, Lambeth Palace Library (London), and Liverpool University Library. Our grateful thanks are due to Dr. J. V. Field who read the entire commentary and made valuable suggestions. The *Praxis* relied less on formal proof than on the immediate evidence of the equations arranged on the page. We have taken pains to preserve the visual impact of the *Praxis* — and this would not have been possible without William Adams' expertise in LaTeX. In particular, he typeset the most challenging part of the *Praxis*, the Numerical Exegesis.

At an early stage of the project, Mordechai Feingold offered invaluable advice and support, and encouraged us to submit our manuscript to Springer for publication in the series in which it now appears. This turned out to be an excellent fit for our book, and we are also grateful to the series editor Jed Buchwald and to Mark Spencer, our editor at Springer.

Above all, we would like to express our thanks to the British Society for the History of Mathematics and the Harriot Seminar. Each provided a generous grant which not only helped financially but was also valuable as a moral support for our work.

Table of Contents

Preface   v

Introduction   1

The Practice of the Analytic Art (translation) 17

Preface to Analysts 19

Definitions 23

Section One 31

Section Two 35

Section Three 49

Section Four 71

Section Five 95

Section Six 105

Numerical Exegesis 131

Rules for Guidance 183

Commentary 209

    Notes on Preface to Analysts 209

    Notes on Definitions 209

    Notes on Section One 213

    Notes on Section Two 217

    Notes on Section Three 223

    Notes on Section Four 229

    Notes on Section Five 233

Notes on Section Six 239

Notes on Numerical Exegesis...................................... 253

Comparative Table of Equations Solved.............................. 263

Textual Emendations 271

Appendix.. 279

Select Bibliography... 293

Index to Introduction and Commentary.............................. 295

Introduction

Revolution in mathematics means the birth of the new but not the demise of the old, only its obsolescence. And in mathematics, as in any other field, the particular aspects undergoing such change must be specified: for example, symbolism, methodology, type of problem, method of proof, axiomatic structure, level of abstraction, or, perhaps, methods of computation.

For reason of brevity, I assume the possibility of a model for the history of mathematics in western Europe, the defining characteristic of which is progressive abstraction. There may be others. I would argue that in the sixteenth and early seventeenth centuries, algebra underwent changes that involved genuine novelty, a revolution one might say, rendering previous assumptions, symbolisms, methodology, goals, and so on, or any combination of these, obsolescent but not invalidating them. And Thomas Harriot (c. 1560–1621) undeniably, played a considerable role in this transformative process.

In the algebraic work of Thomas Harriot, it was above all his notation that was revolutionary. His algebra was the first to be totally expressed in a purely symbolic notation (traditionally, using letters and operational signs), and this was the case in both his manuscripts and in the work published under his name as *Artis analyticae praxis* (1631, London). There appears in his work for the first time ever, the possibility of algebraic logic embodied in the very notation itself, which renders such logic manifest. In Harriot's algebra, we can check that the symbolic manipulation obeys the rules for manipulating algebraic quantities set out at the beginning of the *Praxis* (pp. 11). The rigour for so long associated only with Euclidian geometry now has a new field of operation—algebra.

Yet, Harriot is known in general histories of mathematics principally for certain technical innovations in algebra—for the invention of the inequality signs, for equating the terms of a polynomial equation to zero, and for generating such equations from the product of binomial factors, thereby displaying their structure. It is only since the late 1960s and 1970s with the work of R. C. H. Tanner, Jon Pepper, D. T. Whiteside, and others, that his work has received serious and scholarly attention.

Undoubtedly, his work in algebra was overshadowed in its own time by that of Descartes (1596–1650), whose *La Géométrie*, published only six years after Harriot's posthumous work, would go beyond that of Harriot in achievement and potentiality for future development, but nothing can diminish the credit due to Harriot for his own achievement. It is the contribution that his book made to the ongoing revolution in mathematics of the late Renaissance that justifies the publication, for the first time, of an English translation of the *Artis analyticae praxis*, making it more accessible to modern readers. Such a translation is, in our view, long overdue.

Thomas Harriot was born into a world in which traditional ideas were under intense challenge. Dee's pupil, Thomas Digges (1546–1595), was the first Englishman to publicize Copernicanism and did this in the vernacular. His father, Leonard Digges (c. 1520–1559), advocated teaching mathematics to artisans. There was in fact considerable rapport between the leading scholarly mathematicians and mathematical practitioners in the England of that time. The fact that Digges is published in English suggests a relatively high level of literacy.

The economic context for this collaboration was the rapid emergence of English mercantile capitalism (or, perhaps, imperialism). Dee was technical advisor to the Muscovy Company as Harriot was later to be a member of the Virginia Company. Harriot was friendly with Dee and Hakluyt, corresponded with Kepler (1571–1630) on optics and the telescope, and even made telescopes for sale during the final twelve years of his life. We cannot do better than quote D. T. Whiteside's summary of his accomplishments.

"Harriot in fact possessed a depth and variety of technical expertise which gives him good title to have been England's—Britain's—greatest mathematical scientist before Newton. In mathematics itself he was the master equally of the classical synthetic methods of the Greek geometers Euclid, Apollonius, Archimedes and Pappus, and of the recent algebraic analysis of Cardano, Bombelli, Steven and Viète. In optics he departed from Alhazen, Witelo and Della Porta to make first discovery of the sine-law of refraction at an interface, deriving an exact, quantitative theory of the rainbow, and also came to found his physical explanation of such phenomena upon a sophisticated atomic substratum. In mechanics he went some way to developing a viable notation of rectilinear impact, and adapted the measure of uniform deceleration elaborated by such medieval 'calculators' as Heytesbury and Alvarus Thomas correctly to deduce that the ballistic path of a projectile travelling under gravity and a unidirectional resistance effectively proportional to speed is a titled parabola—this years before Galileo had begun to examine the simple dynamics of unresisted free fall. In astronomy he was as accurate, resourceful and assiduous an observer through his telescopic 'trunks'—even anticipating Galileo in pointing them to the Moon—as he was knowledgeable in conventional Copernican theory and wise to the nuances of Kepler's more radical hypotheses of celestial motion in focal elliptical orbits. He further applied his technical expertise to improving the theory and practice of maritime navigation; determined the specific gravities and optical dispersions of a wide variety of liquids and some solids; and otherwise busied himself with such more conventional occupations of the Renaissance *savant* as making alchemical experiment and creating an improved system of 'secret' writing". [*Hist. Sci.*, xiii (1975), (61–70)]

Life and Reputation

Thomas Harriot is thought to have been born about 1560, probably near Oxford. He certainly graduated from St Mary's Hall, Oxford, part of Oriel College, around 1580, after which he entered the service of Walter Ralegh, who remained his patron until the 1590s.

Harriot had been sent by Ralegh on Sir Richard Grenville's expedition to Virginia (the territory in question is now North Carolina) in 1585 as surveyor, and subsequently published *A briefe and true report of the new found land of Virginia*. This was to be his only publication during his lifetime.

In the early 1590s, Henry Percy, the ninth Earl of Northumberland, became Harriot's patron. This association lasted his lifetime, outliving Percy's imprisonment following the Gunpowder Plot (1605–1621). This membership of the Earl's household brought Harriot into contact with Walter Warner (c. 1557–1643), Robert Hues (1553–1632), and Nathaniel Torporley (1564–1632).

Nathaniel Torporley, a cleric and very reputable mathematician of his time, had considerable admiration for Harriot and was to play a part in the publication of the *Praxis* on Harriot's death. Hues published *a Treatise on Globes* in 1594, was associated with Gresham College, and was appointed by Harriot in his Will to oversee the pricing of his books when they should come to be sold.

The details of the dispersal and disposal of Harriot's papers after his death have been studied and treated in detail in the secondary literature.[1]

Harriot's Will gave Torporley the task of editing his mathematical writings asking him to "peruse and order and to separate the chief of them from my waste papers, to the end that after he doth understand them he may make use in penning such doctrine that belongs unto them for public uses...." If Torporley did not understand the notation, he was to confer with Warner or Hues. Failing this, Protheroe and Aylesbury should be asked to help (Tanner, *History of Science, 6,* 1967, p. 5). Finally, after this, the papers should be put into Percy's library and the key to the trunk holding them should be held by Henry Percy, the ninth Earl of Northumberland, Harriot's patron at the time.

---

[1] Rosalind C. H. Tanner, "Thomas Harriot as Mathematician: a Legacy of Hearsay". *Physis* IX, 1967, pp. 235–256, 257–292.

Rosalind C. H. Tanner, "The Study of Thomas Harriot's Manuscripts", I. Harriot's Will, *History of Science, 6,* 1967, pp. 1–16.

Jon V. Pepper, "The Study of Thomas Harriot's Manuscripts", II. Harriot's unpublished papers, *History of Science, 6,* 1967, pp. 17–40.

R. Cecilia H. Tanner, "Nathaniel Torporley and the Harriot Manuscripts", *Annals of Science,* 25, 1969, pp. 339–349.

Jacqueline A. Stedall, "Rob'd of Glories: the Posthumous Misfortunes of Thomas Harriot and his Algebra", *Arch. Hist. Exact. Sci. 54* (2000), pp. 455–497.

Jacqueline A. Stedall, *The Greate Inventions of Algebra: Thomas Harriot's Treatise on Equations*, Oxford University Press, 2003.

Torporley's copies of some of Harriot's mathematical papers, the *Congestor analiticus* and a compilation called the *Summary* (see J. Stedall, 2003, pp. 19–20, 24–26), are in Lambeth Palace Library, London, having been transferred in 1996 from the library of Sion College where Torporley spent his final years. The *Congestor Analyticus* is Torporley's incomplete attempt at a treatise and the *Summary* of what Torporley thought an edition of Harriot's algebra should contain. After the publication of the *Praxis*, Torporley wrote a criticism of the work, entitled: *Corrector analyticus artis posthumae Thomae Harrioti*. Harriot's own papers, mathematical and others, are in various libraries, but those that are relevant to the contents of the *Praxis* are in the British Library under the catalog numbers Add. MSS 6782-9, almost all in 6782-4. The pages that deal with algebra constitute only a fraction of this material, and the pages that are certainly relatable to the *Praxis* (concerned with the theory of equations and the solution of polynomial equations with numerical coefficients) are interspersed with pages on other mathematical and scientific topics and some sheets that are undoubtedly "waste."

Torporley planned, but never completed, the task assigned to him by Harriot. Within a few years of Harriot's death, the work appears to have been shared between Torporley and Warner. Whatever notes Warner used to put together the *Praxis*, whether he worked from the manuscript papers or a later draft by Harriot, now lost, he certainly interfered with what Harriot had intended and rearranged, altered, reduced, and augmented, as we know from Torporley's copy of Harriot's notes. In a recent study, Jacqueline Stedall has discussed in considerable depth the relationship of the text of the *Praxis* to Harriot's surviving manuscripts in the light of Torporley's *Congestor* and *Summary*. Although the present authors do not fully concur with Stedall's arguments, it is clear that her book presents a "treatise"as close as is possible to what Harriot might have published had he lived to do so (J. Stedall, *The Greate Invention of Algebra*, OUP, 2003).

In the end, it was Warner who edited Harriot's papers and redrafted and rearranged them for the *Praxis*. Whether he did this from the papers now in the British Library, from the Torporley papers or from a further re-working by Harriot, or from other copies, we cannot tell. The history of the papers in the few centuries after Harriot's death has been speculatively traced by several writers, notably R. C. H. Tanner [see *Annals of Science*, 25, 1969, pp. 339–49].

The influence of Harriot's mathematical writings on later English mathematicians such as John Pell (1611–1685), Charles Cavendish (1592–1664), and (indirectly upon) John Wallis (1616–1703) was considerable. Certainly, he had the highest reputation among his contemporaries. Warner's commendation of Harriot took the form of saying: "Harriot... ought truly to be considered to have completely perfected numerical Exegesis, an art which is instrumental in all Mathematical arts and on that account the most useful" (*Praxis*, Preface, p. 4).

In what sense was Harriot's art new? Principally for the much more convenient and practical character of the numerical exegesis previously presented by Viète. The specific means that Harriot used were "a literal notation: that is, the letters of the alphabet, either by themselves or in any combination, according to the needs of the calculation of the reasoning" (*Praxis*, Preface, p. 4). This (says Warner) is to be compared with the work of Viète, who had proposed a "logistic which was to be exercised through (verbally) interpreted signs; although this perhaps was useful for understanding the new discipline, it was subsequently found to be inconvenient for normal practical use." Hence, the first object of Warner's praise was Harriot's symbolism. Processes hitherto "somewhat irksome and rather ungainly" had "been bought to the utmost simplicity and lucidity. (For a description of Viète's notation, see below, p. xx).

Warner next gives the reader what he considers to be Harriot's two chief discoveries (which Warner sees as relying on what he calls "the dexterity of this *Arithmetic*"): first, the generation from binomial roots of Canonical Equations so that, when they are "applied to" common equations, the roots of the latter are revealed. ("...a most ingenious discovery," writes Warner); second the derivation of Canonical Polynomials.[2]

The "uniform and continuous" application of certain rules or Canons enables the mathematician to work with ease and certainty. The latter, thinks Warner, is Harriot's most important discovery and a mere glance at corresponding cases in the *Numerical Exegesis* (below) and Harriot's manuscripts, on the one hand, and Viète's *De numerorum potestatum resolutione* (1600), on the other hand, makes this quite clear to us. Hence, Harriot was a master mathematician and Analyst in the opinion of his editor, Walter Warner. And Warner was not alone in his high

---

[2] A Canonical Equation was one expressed in a standard form produced by the multiplication of binomial factors. What Harriot did may be summarised (for a quadratic equation) in modern notation, as follows:

If a is to be a root of the equation,		$x = a$,
	then	$x - a = 0$
If b is a root,		$x = b$
	then	$x - b = 0$
It follows that		$(x - a)(x - b) = 0$
	i.e.,	$x^2 - (a + b)x + ab = 0$
	and	$x^2 - (a + b)x = -ab$

The expression in the final line is the Canonical Equation and it follows that any equation in such a form has roots a, b.

A Canonical Polynomial may be explained (again, in modern notation) as follows: Problem 2 of the Numerical Exegesis solves an equation of the form:

$$a^2 + da = f^2$$

in the particular case:

$$a^2 + 432a = 13584208$$

Put	$a = b + c$
Then	$(b + c)^2 + d(b + c) = f^2$
i.e.	$b^2 + 2bc + c^2 + db + dc = f^2$
i.e.	$db + b^2 + dc + 2bc + c^2 = f^2$

The Canonical Polynomials are $db + b^2$ and $dc + 2bc + c^2$, which are used in tabular fashion to perform the computation. (See n.2 on Numerical Exegesis).

opinion of Harriot. William Lower, a friend of Harriot's is well-known for having urged Harriot to publish (W. Lower, from J. W. Shirley, *Thomas Harriot*, Oxford, 1983, p. 400).

Despite the waning in Harriot's influence in the eighteenth and nineteenth centuries, he was always remembered. In 1777, Lagrange (1736–1813) credited Harriot with having been the first to show the conditions for a cubic equation lacking the second term to have real roots (Tanner, *Physis*, 1967, p. 283). In 1883, a letter from the algebraist J. J. Sylvester (1814–1897) to Arthur Cayley (1821–1895) referred to Harriot: "It was gratifying, however, to see the handwriting of the man who first introduced the Algebraic Zero into Analysis, the father of current Algebra" (John Fauvel, *Harrioteer*, September 1996). A full appreciation of his work is not possible without considering its historical context, which, in turn, leads naturally into a discussion of the *Praxis* itself.

Historical Background

The second half of the sixteenth and the first half of the seventeenth centuries saw the burgeoning of new methodologies, notations, objects, and relationships in algebra, accompanied by a movement toward generality. Over and above this came a small move toward extending and generalizing the concept of number with Cardano's acknowledgment of negative roots in *Ars Magna* (1545). The "irreducible" case was noted, that is, the case of the cubic equation with three real roots, two of which are reached via conjugate complex numbers. But, although Cardano solved the cubic by a "recipe" resembling those used by the Babylonians (with only an implicit formula), his "demonstration" (a form of proof) of the solution was a geometric one. In Cardano there is no overt demonstration of the algebraic logic leading to the solution. Algebraic logic is not the same as (Aristotelian) formal logic, but it does include it. Manipulation of the real numbers is governed by distributive, commutative, and associative laws, which are axiomatic, but the inferential links are those of formal logic.

Viète is the first algebraist to use a literal sign for a general number, and algebra becomes the "ars analytice," which has a geometrical aspect due to its being related to magnitude.

In the course of the Renaissance, a number of different notations had emerged, to a large extent embodying different concepts, but four principal strands may be distinguished. All of them were in use in the first half of the seventeenth century.

"Cossic" numbers were used by Luca Pacioli (1445–1514) in the *Summa de arithmetica, geometrica, proportioni et proportionalita* of 1494 and were used by Michael Stifel (c. 1487–1567) in the *Arithmetica Integra* of 1544. The latter notation uses single signs (taken from Old German script) for the unknown and each of its powers, e.g., ӿ corresponds to x, ӡ corresponds to x^2, ᴣ corresponds to x^3, and ӡӡ to x^4. This notation pays its respects to the three-dimensional constraint inherited from the Greeks. There is no common reducibility in the Cossic sign-system, just as there was none in Diophantus (i.e., x^2, x^3 have a common "base" absent in the Cossic system). Moreover, there is no exponent acting as an operator (as in x^2).

A second strand, going back to Nicolas Chuquet (c. 1500) and Rafael Bombelli (c. 1526–1573), is embodied principally in Simon Stevin (1548–1620) and Albert Girard (1590–1663). Hindu–Arabic numerals play the decisive role in this tradition, acting as coefficients and as indicators of the powers of numbers, if not exponents in the modern manner. Stevin, in the *Arithmetic*, uses numerals inside circles: ①, ②, ③ represent (our) x, x^2, and x^3, respectively.

It would be a mistake to see this notation as "lacking the unknown," at least subjectively for those using it. Certainly, this notation has a strong bias toward an operational exponent and a uniform quantitative base rooted in unity conceived as a number.

The third strand, found as early as Michael Stifel (1486–1567) and culminating first in Harriot and then in Descartes' *La Géométrie* (1637), ultimately carried the day, becoming standard traditional algebraic notation. In its earliest form (1, 1A, 1AA, 1AAA...) (M. Stifel, *Die Coss Christoffs Rudolfs*, V, 1553) no exponent is overtly used, but from the start it represents a significant departure from the Cossic and other geometrically expressed systems. This form of notation is used by Harriot, although Descartes in his correspondence used a wide variety of notations as the fancy took him or, perhaps, according to the person to whom he was writing.

The fourth strand is represented by François Viète (1540–1603) whose notation in all works, apart from the *De numerosa potestatum resolutione* (1600), in which numerical equations are solved, appears thus:

Sit data B differentia duorum laterum, & datum quoque D adgregatum eorumdem. Oportet inuenire latera.

Latus minus esto A, maius igitur erit $A + B$. Adgregatum ideo laterum A bis $+$ B. At idem datum est D. Quare A bis $+$ B aequatur D. et per antithesim, A bis aequabitur $D - B$, & omnibus subduplatis, A aequabitur D semissi, minùs B semisse.

Vel, latus maius esto E. Minus igitur erit $E - B$. Adgregatum ideo laterum. E bis, minùs B. At idem datum est D. Quare E bis minus B aequabitur D. & per antithesim, E bis aequabitur $D + B$, & omnibus subduplatis E aequabitur D semissi, plùs B semisse.

This is in the 1600 edition of the Viète work, which was followed in 1646 by its appearance as part of *Opera Mathematica*, edited by F. Schooten (1615–1660 or 61). The sign $-$ is used in 1600 as well as the word minus for the subtractive operation, and whenever the word is used thus, it is changed by Schooten to $-$ in 1646. *De Numerosa Potestatum Resolutione* (1600) uses $-$ throughout, although the sign is usually slightly broken due to the exigencies of printing. The word "bis," used in 1600 is changed to 2 by Schooten, i.e., A bis becomes A2. "D semissi plus B semissi" in the above extract becomes $D^{1}/_{2} + B^{1}/_{2}$ in 1646 and "D semissi minus B semissi" becomes $D^{1}/_{2} - B^{1}/_{2}$ in 1646, as a result of Schooten's editorship.

In Viète's logistica speciosa, "A quadratum" corresponds to, but is not to be identified with, A-squared (A^2). "A quadratum" denotes "a square," the "side" of which is A, following the Greek tradition. A, B, C, etc. were known as "species".

"A quadratum" is a development upon Diophantus' square number denoted by the first two letters of δυναμισ (dunamis). Harriot would write simply *aa*.

François Viète (1540–1603) saw himself as "restoring" the mathematics of the classical world and saw his *Introduction to the art of analysis* (*In artem analyticen isagoge*), as part of the *restoration of mathematical analysis* (*opus restitutae mathematicae analyseos*). However, despite his subjective view of his own project, what emerges objectively in his work is a decisive transformation in which Analysis (and Synthesis) will, for the first time, be identified with algebra as well as geometry. The title of Viète's book varies in different editions. Those of 1591, 1624, and 1631 have *In artem analyticum isagoge*, the 1635 edition has *In artem analyticam isagoge* and that of 1646 has *In artem analyticum isagoge* (Witmer, 1983, p. 11).

Perhaps Viète's most outstanding contribution to algebra lies in the new level of generality engendered by his notation in which, for example, A represents a general positive number. On the other hand, the link with a geometrical base is never broken and even *On the Numerical Solution of Equations* is followed by *Canonical Rescension of Geometrical Constructions* (Tours, 1593), which (as Mahoney wrote) "showed how symbolic quantities may be interpreted as line segments and symbolic operations as geometric construction procedures" (Mahoney, 1994, p. 38).

It is, however, of more importance that it is made clear in the *Isagoge*, especially in Chapters III and IV, that the entities being referred to are magnitudes, and this is so in all of Viète's work except for the (numerical) solution of equations with numerical coefficients. Magnitudes are, in fact, geometrical quantities, lengths, areas, or volumes. Such a magnitudinal base accounts for the retention by Viète of Greek geometrical constraints. Chapter III opens: "The prime and perpetual law of equations or proportions which, since it deals with their homogeneity, is called 'the law of homogeneous terms' is this: 'Homogeneous terms must be compared with homogeneous terms' " (Witmer, 1983, p. 15). Viète's algebra is thus both geometrically and numerically based and he builds on the work of Pappus and Diophantus.

Diophantus had flourished c.250 AD and Pappus a half a century later. The first provided Viète with a basis in (arithmetical) algebra and the analytic solution of algebraic problems and Pappus provided geometrical analysis. Viète used both these classical authors and subsumed them into his new analytic art. He did not, however, see himself as an innovator but as a renewer of the classical tradition.

Viète and Harriot

Unlike Viète, Harriot did not base his notation upon magnitude even though he applied algebra to geometrical problems. (See, for example, British Library Add. MS 6784, ff. 19–28.) For both authors, however, the distinction between "Numerical" and "Specious" Logistic (Arithmetic) is given as that between representation by numerical signs or by letters of the alphabet. The first mention of *logistice speciosa* by Viète is in Chapter I of *In artem analyticen isagoge*: "It no longer limits its reasoning to numbers, a shortcoming of the old analysts, but works with a newly discovered symbolic logistic..." (Witmer, 1983, p. 13). The full definition by Viète comes at the beginning of Chapter 4. Numerical logistic is [a logistic] that employs numbers, symbolic logistic is one that employs symbols or signs for the things as, say, the letters of the alphabet (Witmer, 1983, p. 17). Such wording accords very well with the practice of Harriot and accords with Definition 1 of the *Praxis*. It would also accord with the later views of Wallis, as expressed in his *Algebra* (1685, p. 12). (See Witmer, 1983, p. 13, fn. 8 for full discussion.)

Harriot studied Viète closely and was affected by him both instrumentally and conceptually. In the manuscripts, Harriot's solutions of polynomial equations with numerical coefficients are all accompanied by a reference to the corresponding solution by Viète and, moreover, all the pages are headed *De numerosa potestatum resolutione* (*On the Numerical Resolution of Powers*) the title of Viète's own published work (1600). Harriot's debt to Viète is clear in the algorithmic outline of the solution. The Definitions demonstrate a conceptual debt to Viète but, importantly, the manner in which his notation transcends that of Viète.

Harriot's algebra was expressed in a completely symbolic notation unlike that of Viète, which had included linguistic elements. In a sense, Harriot's algebra by-passed the species of Viète, which had their roots in Diophantus and were bound to Greek ideas about the necessity for homogeneity. Homogeneity was retained throughout the Harriot manuscripts, as it was in the *Praxis* except for the cases in which a collection of terms was equated to zero.

It was Descartes (who said that he had not read Viète before writing *La Géométrie*) who dealt with the issue of homogeneity by the use of "dummy" units. As early as the *Regulae ad directionem ingenii, Rules for Right direction of the mind* (posth. 1684), Descartes pointed out in Rule XV the need for an arbitrary unit and followed this up in Rule XVIII in connection with a problem involving continued proportion. The Rule continues with an "algebra of magnitudes" in which multiplication (product) is defined by taking two magnitudes a and b and asking the reader to imagine them as the sides of a rectangle. Then, "if we wish to multiply ab by $c,...$ we conceive ab as a line, viz. ab."

Descartes had already distinguished conceiving from imagining in Rule XIV. Imagination involves a mental image but conception is a faculty of understanding that requires us only "to attend to what is signified by the name". Thus, he has given meaning to the product ab as a line by postulating it as a theoretical construct but having a basis in what can be visualized. Hence, even in the *Regulae* Descartes had confronted this issue.

Things are dealt with differently in the *Gèomètrie*, where unity is used directly to dispose of the problem of homogeneity. First, he uses proportions in similar triangles, having taken one side as unity, and simply shows that another side is equal to the product of the remaining sides. A little further on, the issue is tackled algebraically when he considers $a^2 \ b^2 - b$ and we must consider the first term divided by unity once and the second quantity multiplied twice by unity. The important thing is that Rule XV of the *Regulae* had said of unity that it is "an object extended in every direction and admitting of countless dimensions". So, in effect, each "dimension of unity" can deal with a different magnitude of dimension.

Harriot did none of this (as far as we know) and homogeneity is maintained in the manuscripts even so far as showing the product of eight zeros (Add. MS 6783, f. 187). For Harriot, however, letters stand for particular unknown numbers (and occasionally lengths) and the need for a unity of Descartes' type just does not arise.

Descartes bases his algebraic geometry on a combination of magnitude and number. Magnitudes are numerically defined via the letters of their names that also stand for numbers, for example, the magnitude a is a length designated by "a," which is a general number and used algebraically. It is the combination of this with their use in a locus problem implicitly involving motion that will enable curves to be expressed by equations and facilitate the emergence of the real variable. Harriot's notation satisfies the necessary conditions for an algebraic symbolism but is insufficient, because of the superfluity of unknowns arising from the homogeneity requirement, unlike the algebra that follows from the usage of Descartes.

The Preface and Definitions of the *Praxis* are unique in having no corresponding pages in the Harriot manuscripts. We do not know, therefore, whether the Definitions are the unaided work of Warner or whether he was using a draft by Harriot, which has since been lost. The major interest in the Definitions lies in the first twelve, which refer to such terms as Logistice Speciosa, Analysis, Synthesis, Zetetic, Poristic, and Exegetic, all of which had been used by Viète. An assessment may be made of the conceptual connections between Harriot and Viète by a study of these Definitions, but this is so only if Warner's version can in some way be identified with Harriot's. No trace of the Definitions has come to light in any of Warner's papers, nor in Torporley's copy at Lambeth Palace, and as we have already remarked, nor is it to be found in the Harriot manuscripts themselves. However, Warner knew Viète's *Isagoge* and Add. MS 4394, f. 108 has a table, in the left-hand column of which is a list of mathematical works and in each corresponding place in the right-hand column is a name. Against Viète's *Isagoge* is the name Warner and, although there is no hint of the nature of the connection (ownership?), it is clear that some form of association is attested to by this.

Pages in Harriot's manuscripts, particularly Add. MS 6784 ff. 19–28 (not used in the *Praxis*) show all the above terms (albeit with reference to work not directly relevant to the *Praxis* itself). There is, therefore, sufficient evidence to permit us to connect Harriot and Warner conceptually with Viète. A detailed comparison of the Definitions with appropriate passages appears in the textual comments on the Definitions (see below). It is clear that, in the reference to Logistice Speciosa in the *Praxis* in Definition One, what is meant by Speciosa (specious) is the use of letters rather than numbers and Harriot favors a notation in which a letter of the alphabet represents a number seen as composed of numerical units. It will be seen below that the word was used simply to designate literal notation by Warner, Harriot's editor. Furthermore, it will be seen that analysis and synthesis (analysis in the three forms mentioned by Viète, Zetetic, Exegetic, and Poristic) all become exclusively algebraic in Harriot's hands in the manuscripts and in the *Praxis*. It also becomes clear that different sections of the *Praxis* correspond to each of the processes. What, then, are the contents and achievements of the *Praxis*?

Contents and Achievements of the *Artis Analyticae Praxis*

As the Contents page indicates, the book is in two parts, of which the first, up to the end of Section 6, is on theory of equations and the second, the Numerical Exegesis, is given over to the solution of polynomial equations with numerical coefficients. The *Preface* provides a broad outline history of Analysis. In fact, the historical background is used by Warner to provide a backdrop for a eulogy to Harriot, seen as "a new Viète," who is himself credited with the invention of numerical Exegesis and "specious" (i.e., in letters) Arithmetic. However, Viète's logistic is found by the author to be "inconvenient for normal practical use" and it is Harriot's (purely) literal notation which remedies this and brings Analysis to "the utmost simplicity and lucidity."

Moreover, it is not surprising that the polynomials are especially welcomed, providing as they do a general, algorithmic, purely calculative and non-verbal method of solution for equations with numerical coefficients by successive approximation. This method may now be outdated, but such algorithmic rules constitute a watershed between the situation when verbal directions were written in prose to describe a procedure and the framing of such directions in a purely quantitative form like a calculation in arithmetic.

Such polynomials were, according to Warner, Harriot's second major discovery. The first had been to generate Canonical equations from "binomial rootes" (that is, factors, see Section Two, pp. 12–28, *Praxis*) so that the roots of common equations may be revealed by comparison. Such determination of roots would be done in Section 5 of the *Praxis* for which the manuscript evidence, apart from the inequalities, is extremely sketchy. Yet, Warner states this as Harriot's achievement with confidence and some evidence appears in the Torporley papers (now in Lambeth Palace Library but previously in Syon House, Arc. L.40.2 L.40), which were taken from the Harriot manuscripts and are really abbreviated notes.

On 44v and 45r there is some algebra, which is difficult to decipher, accompanied by an inequality replicating an inequality in Section Five of the *Praxis*:

$$\frac{qqr+qrr}{2} \;\Bigg|\; < \; \frac{qq+qr+rr}{3}$$
$$\frac{qqr+qrr}{2} \;\Bigg|\; \frac{qq+qr+rr}{3}$$
$$\frac{qq+qr+rr}{3}$$

There seems to be some comparison of equations but no determination of roots can be seen. It follows that either certain manuscript papers have been lost or that Torporley did have access to a draft or set of notes which went further than Harriot's work in the manuscripts.

We may also make another deduction. It is in the *Preface* that one might have expected a reference to the omission of non-positive roots in the work. But the author writes, "any uncertainty of the roots ... is revealed and dispelled". Yet hundreds of cases exist in the manuscript papers in which negative (and occasionally also imaginary) roots are given. It must be noted that in almost all manuscript pages, negative roots are recognized. There are about 400 equations, quadratic, cubic, and biquadratic, all with numerical coefficients, displayed systematically in lists and accompanied by a simple statement of their roots, but without algebraic derivation. In only five cases are negative roots not given and these are imaginary.

Cases exist, however, of positive and negative imaginaries AD MS 6783, f. 49, f. 156, and f. 301 that display negative imaginary and complex roots and see p. 221 for a summary list of different roots, dealing with all equations shown in the manuscripts apart from those pages which are obvious waste.

Warner takes no account of this. Perhaps he had never seen the pages with negative (and complex) roots? Perhaps Warner did not recognize them? One wonders whether such issues may not have been discussed between Warner and Harriot. And what were Harriot's own views? The disparity between manuscript pages in which non-positive roots were given and those in which they were ruled out suggests the possibility that Harriot's views changed in the course of his life. It is unlikely that different papers were intended for different audiences since he goes to the trouble to "prove" by substitution in his papers on the Generation of Canonical Equations that the positive roots are the only possible ones. Moreover, factors corresponding to negative roots are meticulously avoided. Especially noteworthy is the page showing a biquadratic equation solved completely, in which all four roots are given (two real and two complex) in P. Rigaud, *Miscellaneous Works and Correspondence of the Rev. James Bradley, D.D., F.R.S., 2 vols. (Oxford, 1832–3). Supplement—with an account of Harriot's astronomical papers, Plate V.*

Warner must surely be given the benefit of the doubt as regards his motives and must be assumed as having acted, however mistakenly (and perhaps, incompetently), in good faith. The only other alternative is to suppose him to have been guilty of perverse falsification. And we must bear in mind that the Torporley copy itself followed the manuscript pages in precluding negative roots. Hence, it may very well have been Harriot's own intention to ignore negative roots, such intention springing from his belief at the time or for pedagogical reasons. (Or, perhaps, he just did not want to stir up controversy?)

The Definitions have been touched upon above (see p. 9, above) and are dealt with in detail below. They are given, as the author indicates, to make clear the precise meanings of terms in common use. We might speculate, however, that a further reason for their inclusion is to provide Euclidean "credentials" for the work as a whole, a practice which was not at all uncommon then and for a considerable time to come.

Section 1 gives lists of examples of the four rules and some rules for the manipulation of equations. In Section 2, binomial factors are multiplied to produce (with a little manipulation) Primary Canonical Equations, and this leads straight into Section 3 in which one or two terms are removed from such equations in order to reduce them to what are called Secondary Canonical Equations, which are in a form more suitable for solving. The removal of one term is achieved simply by equating the appropriate coefficient to zero. However, the removal of two terms creates difficulties for Warner (see below, pp. 46, *Praxis*). The results obtained in Sections 2 and 3 are then used in Section 4 for the designation of (positive) roots as a preparation for Section 5. In this section, what are called common equations, are compared with the Canonicals and, on the basis of the results of Section 4, conclusions are drawn on the numbers of their roots.

The final Section of the theoretical part of the book removes the next-to-highest term from polynomial equations (cubic and biquadratic) following the method initiated by Cardano and followed by Viète and, in the cases of Problems 12 and 13, actually solves the equations. However, Section 6 is exceedingly repetitious, using the same method time after time and we can only suppose that the intention was to provide practice for the reader or to present all possible cases, however repetitious.

Up to Section 5 there seems to be a rationale for the ordering of the sections. It is only with Section 6 that this seems not to be the case. Why should such a section be inserted after Section 5 and before the Numerical Exegesis when it has no immediate connection with either? It can be shown that Warner rearranged the material he was working on in a way which contradicted Harriot's intentions (see below Appendix, p. 277) resulting in a structure which was not wholly coherent. Warner's rearrangement explains the apparent inappropriateness of the structure of the *Praxis*.

The "practical" part comes next, the numerical solution of equations by the method of successive approximation. The method is distinguished from all previous presentations by the use of polynomials, written out in lists in a beautifully clear symbolism designed for computation by the reader, which are called by Warner, Rules for Guidance. As the *Preface* and the final words of Section 6 had explained, this section is the principal object of the whole work.

The particular achievements ("discoveries" as the *Preface* calls them) all refer to and are dependent on the new notation. Multiplication is denoted by $\lrcorner$ and when the binomial factors are multiplied to generate equations, all is very clear on the basis of the notation:

$$(a - b)(a - c) = a^2 - (b + c)a + bc \text{ is written } \begin{array}{c} a - b \\ a - c \end{array} \Big| === \begin{array}{c} aa - ba \\ -ca + bc \end{array}$$

Here, the work is set out as in simple arithmetic. Christopher Clavius (1537–1612) had done something similar, but the Cossic notation in which he expressed everything did not lend itself to being directly related to simple arithmetic. Harriot has been credited with being the first to equate all the terms of an equation to zero. He may have been able to do this because there is less likelihood of seeing the side "opposite" to the zero as "something" equated to "nothing" when the entire equation is seen abstractly and symbolically. However, Harriot never leaves the equation in this form and a final line is always added equating all to the "homogene" even if this is negative. (Viète called the "constant" the "homogenea comparationis". See *Isagoge*, Chapter V, *On the Rules of Zetetics*.)

Section 5 calls for special mention for it is in Section 5 that a general method is postulated as providing conditions for determining the roots of literal cubic (and one biquadratic) equation(s) based upon the results of Section 4. In several cases, a common equation is compared with a Canonical equation on the basis of their having comparable inequality relations and the number of roots of the former are then determined in terms of those in the Canonical obtained in Section 4.

The idea is a good one but the reasoning is flawed. First, the inequality condition for the identity of the roots is simply asserted without justification. Second, the conditions given are necessary but not sufficient.

Cecily Tanner ("*Thomas Harriot as mathematician: a legacy of hearsay. Part 2*," *Physis* 9, 1967a, p. 283) quotes Lagrange as having written in 1777, "Harriot seems to me to have truly been the first to show directly and analytically that third degree equations without a second term cannot have [all] their roots real unless the cube of one third of the coefficient of the third term, taken with opposite sign, is greater than the square of half of the last term". [Harriot me parait être proprement le premier qui ait démontre d'une manière directe et analytique que les equations du troisième degré sans second terme ne sauraient avoir leurs raciness réelles, à moins que le cube du tiers du coefficient du troisième terme, pris avec une signe contraire, ne soit plus grande que le carré de la moitié du dernier terme.] With these words, Lagrange claims just enough for the *Praxis* and implies by his use of the word "unless" that only the necessary condition is given.

Finally, Edmund Halley, who investigated the matter geometrically ("*On the Numbers and Limits of the Roots of Cubic and Quadratic Equations*" (Phil. Trans. Roy. Soc. (translated), 1686–7, 16, 395–407, esp. 398) concluded from a consideration of the equation

$$z^3 - bz^2 + pz - q = 0,$$

that "Also Prop. 5, Sect 5 of our countryman Harriot's Art Analytica and Prob. 18 of Viète's Numer. Potest. Resol. is hardly founded."

Assessment of the Significance of the Praxis

How must the *Praxis* be judged overall in respect of its content and structure? First, its style is fundamentally Euclidean insofar as it begins with Definitions as if setting out an axiomatic system and this, together with the verbosity so unlike Harriot's own work, alters the very shape from a terse algebraic presentation to a sequence of Propositions presented as if they were theorems. Perhaps Warner did this in order to render the work more acceptable and perhaps the extra words (especially in the Numerical Exegesis) might be thought to make the work more accessible. Also, most mathematical works to hand (particularly Viète) were wordy and in Latin prose but, in the light of the vernacular textbooks of Recorde and in the context of the practical proclivities of Harriot himself, it is possible to speculate that Warner might well have aimed at an aristocratic rather than a plebeian readership (or, perhaps, an international one).

Putting such speculations aside, however, the particular achievements of the *Praxis* were many and clear; the exposure of the structure of polynomial equations through their generation by multiplication of binomial factors; equation of all terms to zero; the display of the coefficients in terms of roots for all to see; the treatment of inequality as operational and the establishment in the Numerical Exegesis of the algorithmic method by means of a transparent notation. For the first time, the focus is on the study of the (quantitative) relational structure of equations.

Harriot did not state what has come to be known as Descartes' Rule of Signs, that is, an equation can have as many true [positive] roots as it contains changes of sign from + to – or from – to +; and as many false [negative] roots as the number of times two + or two – signs are found in succession, nor did he state the general relationship between roots and coefficients and did not assert the equality between the degree of an equation and the number of roots. Although many examples in the manuscripts are solved for the correct number of roots, there are a very large number in which this is not the case, particularly biquadratics with two complex (conjugate) roots. (See MS Add 6783, f. 49.) At various times these omissions have been treated as shortcomings but such value-judgments are inappropriate historically. In the history of mathematics we risk absurdities by not judging work by the standards of its own time.

What is indisputable is that Harriot was first in the field to use a truly symbolic notation, and the *Praxis* is the first published algebraic work to be purely symbolic, even though his retention of homogeneity rendered it unsuitable for modern needs. Symbolism of any kind arises from the need to embody in a visualizable form that which is essentially unvisualizable. We recognize today that a mathematical sign-system should embody only the quantitative aspect of what is being represented symbolically and do this totally. The system should be fully cipherized, embracing unknowns, knowns, and operational signs. Such a system carries with it the quantitative logic of the argument or computation involved.

The lack of exponential notation in Harriot's algebra undoubtedly renders his notation incomplete. Nevertheless, the resulting inconvenience is only minor and the need to count the number of letters, though tedious, is itself a quantitative procedure and may have been a godsend in Harriot's day as it did away with the possibility of ambiguity in Cossic notation when multiplying powers.

In Harriot's work we can see that a new zone of mathematical existence has come into being, self-dependent and detached from the substantive base to which it is, nevertheless, connected. Perhaps, the most powerful characteristic of this new zone is its potential for unlimited representation, which can be known only with hindsight. Harriot's notation is unique, unambiguous, number-based and non-linguistic. Harriot has been rightly deemed to be a Renaissance figure with his multiplicity of interests but he fulfills himself primarily by taking algebra out of the Renaissance and into the modern world. And this is not only in manuscript but in print in the *Praxis* in 1631, six years before Descartes' *Geometry*.

It is necessary to distinguish between any mathematical development in its own right, on the one hand, and its role in the historical process on the other. Any assessment of Harriot must credit him unqualifiedly with his achievements independently of comparison with the towering figure of Descartes.

Whatever Descartes acquired from Harriot, no-one could accuse him of plagiarism, however, if only because his "geometrism" contrasts with the numerically based algebra of Harriot. If he absorbed some of Harriot's ideas and transformed them "in his own image," who would blame him? None of us lives in an intellectual vacuum and if Harriot's thought contributed to that of Descartes, it greatly increases Harriot's stature and the measure of his contribution to the process of historical development of mathematics.

Whatever his faults in carrying out the task he was left with, we cannot but feel gratitude to Warner for actually *doing* it. Without the *Praxis*, would Harriot's reputation have continued? Would anyone have bothered to retrieve his papers? We can never be sure. For this alone, we owe a debt to Warner. Harriot's work was indeed well worth preserving and its being passed on to later mathematicians did make a difference, if only to encourage and stimulate mathematical thinking, especially through familiarity with his (relatively) advanced notation.

THE PRACTICE OF THE ANALYTIC ART

For solving Algebraic equations by a new, convenient
and general method:

A TREATISE

Transcribed with the utmost accuracy and care from the last
papers of THOMAS HARRIOT, the celebrated Philosopher
and Mathematician:

AND DEDICATED TO THE MOST ILLUSTRIOUS LORD
LORD HENRY PERCY
EARL OF NORTHUMBERLAND

Who ordered this work to be newly revised, transcribed,
and published for the general use of Mathematicians—a work
which was first composed for his own use, under the auspices
of his Generosity and Patronage, and therefore a dedication
which is most richly deserved.

LONDON

by ROBERT BARKER, Royal Printer:
And Successor to John Bill, in the Year 1631.

Preface to Analysts

It was the Frenchman FRANÇOIS VIÈTE—an eminent man and, due to his outstanding skill in the mathematical sciences, an ornament to his nation—who first, in a remarkable undertaking and by an unprecedented effort, set about the restitution of the analytic art (the subject of this book) which had long lain ignored and neglected since the learned age of Greece. He has left to posterity indisputable evidence of his noble mental endeavour in the various treatises which he wrote, both elegantly and acutely, to advance this subject. But although he laboured hard to restore the analysis of the ancients—this indeed was the task he set himself—it seems that he passed on to us not so much an analysis restored, as one augmented and embellished by his own discoveries, an analysis new and (one might almost say) entirely his own. This is to put it in general terms, and it must be explained in more detail: I shall show what it was that Viète first did in furthering his goal, so that then it will be possible to gain a better idea of what our most learned author THOMAS HARRIOT (who was the next contender in the analytical arena) afterwards achieved.

And so, to begin at the very beginning: all the ancient practitioners, in seeking the solutions of problems in which the reasoning did not exceed the limits of the quadratic order, generally employed Analysis. This is obvious in practice in their various works, and they themselves also clearly state this to be the case. It is quite certain, then, that the mathematical sciences which we have received from them were enriched with very many additions by the aid of this investigative art. For, by an Analytical process, they first brought the Problem to the stage of solution, so that it was tractable and simple; then, retracing their steps in the Analysis, they constructed a proof synthetically; and finally, they dispensed with the Analysis, and attached this constructed proof to the Problem. But they were only able to do this within the limits of the common Elements,[1] or (as they themselves put it) while they were dealing with a plane locus.[2] But when they had attempted Analysis, and happened to meet with formulas of higher orders (especially cubics), a solution followed less successfully than they had hoped; and so that they would not seem deprived of all resource in their art to furnish a solution in some Geometric form, they would take refuge either in solid loci (by which name is to be understood Conic sections) or what they called linear loci, (for example Spirals, Conchoids, Quadratrices and similar forms of this kind)—necessary supplements, as it were, of a defective art. Those supplements, however, are complicated lines, mechanically described by compound motions; it is quite impossible to make any calculation or reasoning beyond immediate consequences from the assumptions in their original construction.[3] The result of this was that, by calling in the assistance of these aids, the desired solution of the problem had to be devised in a merely technical manner, with the help of the hand and the eye. Thus the analytical capacity of the ancient Greeks in the solution of problems languished during the whole period in which the study and profession of the mathematical arts flourished among them.

But when Greece was at last conquered by barbarian arms and reduced to slavery, the whole learning of Greece passed over into the Arab schools. There, throughout the succeeding period, it was cultivated and developed to a remarkable degree by the studies of an ingenious people. Now, although in the other branches of philosophy many useful discoveries (and some rather more obscure ones too) were made by their skillful investigations and have come down to us; and although the Arabic name of algebra itself, their own coinage, is strong evidence that the study and practice of this art flourished among them (as are the very few writings of theirs which exist in that field); nevertheless, it is DIOPHANTUS the Greek analyst, the last-born of the ancient line of Mathematicians, who alone forbids us from regarding ourselves as indebted to the Arabs either for the actual invention of algebra, or for anything subsequently added to the discoveries of the Greeks which serves to perfect or extend analysis.

And so the analysis of the Greeks has remained in its original form—in the same state of imperfection, that is, in which they themselves had left it—right up to our own times, passing unmodified through the hands of the Arabs. Until, that is, the Italians CARDANO and TARTAGLIA, celebrated mathematicians of a previous age and devoted students of algebra, basing their work on a foundation of geometry (and strongly disputing between themselves for the honour of the discovery), tried to advance the art to the demonstrative solution of equations of the cubic degree by solving some special cases—accurately enough, but in a form rather complicated by binomial roots; and only special cases, because their basis of solution is not general and absolute. After them, others returned their invention to the anvil. Of these, STEVIN the Belgian in his general Arithmetic handled this subject best and most carefully of all. First, he proposed a mode of solution of those equations of the cubic type which, by their nature and primary form, are soluble (inasmuch as their solution can be constructed immediately from a basic substitution). Secondly, he reduced and solved those forms of cubic equations which can, by their very condition, be reduced to primary forms. Thirdly, he also reduced and solved in the same way those biquadratics which could be reduced to primary cubics. But by his tacit exclusion of the remainder—equations of both the Cubic and the Biquadratic type to which these conditions do not apply—he condemned them (the great majority of the total number) as insoluble, much to the detriment of the art. And this was the extent of the progress of this Italian invention, limited not so much by human ignorance as by the nature of the matter itself.

At last, however, Viète appeared, that great architect in analytic studies. But although he employed the aid of Supplements, Recognitions and Angular Sections,[4] and tried everything to overcome, with the engines of his intellect as it were, this stubborn irregularity in the art of analysis, he does not appear to have advanced the subject much beyond the limit reached by his predecessors. Until, that is, after trying geometry in vain, he applied himself to the arithmetical approach and arrived at the happy invention of his numerical exegesis. After this discovery, he was able to state with confidence that haughty and universal problem of his: to solve every problem. For this is the art that nature itself has ordained for the solution of all equations of every degree and form by a general, uniform and infallible method. And since the solutions of problems are ultimately achieved by the solution of equations, when Viète perceived the tremendous power of this exegesis in the solution of equations, he deemed that a universal solution of problems was possible by its means, and so chose to emphasize that problem by stating it in such a magnificent form. The invention of this exegesis is foremost in dignity among all the elements of Viète's projected work of restitution; so, too, let it take first place in the order of this account.

There remains the second of his own inventions, which he introduced into Mathematics under the title of Specious (symbolic) Arithmetic. This is less essentially pertinent to the restitution of Analysis than is numerical Exegesis; yet, because it far outstrips the other in natural priority, and certainly in the generality of its application, it ought not to be less valued. The ancients were at a great disadvantage in not having this specious Arithmetic—as anyone will recognise who has experienced its incredible aptness for handling Mathematical matter succinctly and lucidly, in comparison with the wordy tedium of the ancient style. Since it is agreed that Viète has enriched the art by these two developments—specious Arithmetic and numerical Exegesis (and no trace whatever exists of these in the writings of antiquity)—he deserves, as has been said, to be commended for creating a new art, at least for the most part, rather than for restoring an ancient one.

It is that numerical Exegesis which we offer here, taken from the papers of our Thomas Harriot; not as it was shaped by Viète's first deliberations, but as it was subsequently reshaped by those of Harriot—and reshaped to such an extent, that if Viète seems to have created in some sense a new Analysis by his invention of the Exegesis, then Harriot, by his revision of Exegesis, has produced a new Viète, certainly one which ought to be admired for its new, much more convenient and practical character, as anyone will judge who compares the nature of each when it is put into practice.

But in order to complete this revision of Exegesis, he first had to alter the form of Viète's Logistic as well. For Viète proposed, both by precept and example, a logistic which was to be exercised through (verbally) interpreted signs; although this perhaps was useful for understanding the new discipline, it was subsequently found to be inconvenient for normal practical use.

Consequently Harriot used only a literal notation: that is, the letters of the alphabet, either by themselves or in any combination, according to the needs of the calculation or the reasoning. The numerous examples in the present treatise clearly demonstrate that, by this appropriate change, the practical use of specious Arithmetic, which was formerly somewhat irksome and rather ungainly, has been brought to the utmost simplicity and lucidity. Relying on the dexterity of this Arithmetic, Harriot set about reshaping the Exegesis chiefly through two discoveries of his. First, he set down certain equations generated from Binomial roots, which he calls Canonical. When these are applied to common equations, any uncertainty of the roots which remains in these common equations is revealed and dispelled, through the equivalence of these canonical equations—a most ingenious discovery. Secondly, when he came to the actual application of the numerical Exegesis, he derived certain polynomials from the expressions themselves of the equations which were to be solved; these he also calls Canonical. For, in fact, certain Canons or rules exist within the resolution itself, and by their uniform and continuous application, the process of the work of Analysis is conducted from start to finish with such ease and certainty that Harriot—by the discovery of this device alone, more than all his other discoveries in this field—ought truly to be considered to have completely perfected numerical Exegesis, an art which is instrumental in all Mathematical arts and on that account the most useful. These are almost all of our Author's achievements in his labour to reshape Exegesis; here given only in summary form, but they are explained thoroughly and in detail in the following treatise, to the great benefit of Analysts.

Definitions

Certain definitions, which (in lieu of an introduction) may assist in understanding the terms in common use in the art, as well as those peculiar to the present treatise

DEFINITION 1 [1]

Specious Logistic: This type of Arithmetic is frequently used and is absolutely necessary in these writings on Analysis; it is a sibling of Arithmetic, through participation in the same genus. For Arithmetic is numerical Logistic. The distinction between them goes no further than that signified by their names: in Arithmetic, the quantities of measurable things are expressed and reckoned by characters or figures peculiar to the art, by numerals, as in measurement generally; in the former, however, the quantities themselves are indicated and in every way handled through written signs—the letters of the alphabet, that is—'speciously', as it were (borrowing the term 'specious' from commercial usage). Hence it has received the name Specious.

DEFINITION 2

Equation is used in its common sense for any sort of equality of two or more quantities; but as a special term of this art, it is the clearly determined equality of the sought quantity with some given quantity, when a comparison has been made of one with the other. The part which is sought is a simple or affected (conditioned) power but the part given is commonly called the given homogeneous term of the comparison or equation.

DEFINITION 3 [2]

In propositions of any sort and in drawing up theorems or problems from them scientifically, the best method of proof and an entirely natural way, is that by

---

Bold numbers in the margin refer to the pages of the 1631 printed edition of the *Praxis*. The preface was unpaginated.

which one proceeds from the first principles and elements proper to each subject, composing a chain of logical steps until the proposition is confirmed. Hence it is called the compositive method or, in the terminology of the ancient practitioners, *Synthesis*.

DEFINITION 4 [3]

Yet, in solving problems—especially those encountered by chance—the Arithmetician very often has no recourse to suitable middle stages for proving the proposition; thus, if he takes the natural route of synthesis from first principles and elements of the science, he cannot make any progress in finding and demonstrating a solution by logical means. And so, in such a case of ignorance (which happens almost constantly) necessity teaches him to resort to a reverse procedure, the opposite of the natural one. He makes a start from some unknown and sought quantity which occurs in the problem, as if it were a known and given quantity; he then proceeds by resolving it in a chain of logical steps, until he reaches an equality between the quantity which was taken as given (standing alone, or raised to a power and qualified by coefficients) and some quantity which really is given. If this equality can be found and properly established by such a technique, then, from this equality, the sought quantity itself either will be self-evident, or can be elicited through a further application of the technique; and so the problem can, at length, be solved. And that is the method of resolution which the ancients called by the descriptive term *Analysis*.

DEFINITION 5

2   The words 'composition' and 'resolution' which were introduced in these two definitions are the usual ones Mathematicians employ, by which they signify explicitly, when necessary, the two contrary paths of reasoning used in constructing demonstrations. In the first, one descends from the more simple and less composite to the more composite—that is, by composition—following the natural structure and ordering of the sciences: in other words, from the prior to the posterior. In the second, however, one ascends to the conclusion from the more composite to the less composite and the simple—that is, by resolution—in a reverse order, contrary to the natural one: in other words, from the posterior to the prior. For, the elements and axioms of the sciences, if they are legitimately constituted, ought to be convertible; hence it comes about that those which in nature are antecedents can, in reasoning, be consequents. It is for this reason that logical progress through deductive steps is, of necessity, equally firm and demonstrative in both directions.

DEFINITION 6 [4]

From the above definition of Analysis it may be inferred that it consists of two parts, distinguished by function. The first part is wholly concerned with establishing equalities. As has been said, it begins from the thing which is sought, assuming it as if it were given; then it aims at finding and establishing, by logical consequences, an equality between the sought quantity (which was assumed as known) and some given quantity; and, having established the equality, it terminates and rests there. And since this sort of establishment of equalities consists in the skilled examination of the sought quantity, the ancients called this art *Zetetic*, meaning an 'investigative' or 'inquisitive' art.

DEFINITION 7 [5]

The second part of Analysis has this purpose: from an equality established by Zetetic, the sought quantity is revealed by a continued or a modified type of re-solution: either by a symbol (if it can be revealed by an algebraic expression), or by a numeral (if a numerical solution is required)—and so, at length, a complete solution to the proposed problem is established. François Viète, the great master of the art of Analysis, called this part of Analysis *Exegetic*, meaning 'declaratory' or 'evidential'.

DEFINITION 8 [6]

The ancient Analysts, in addition to Zetetic (which is specifically concerned with the solution of problems) distinguished another kind of Analysis: *Poristic*, mea-ning 'deductive'. This is used to examine whether theorems put forward by chance are true or not. The method of both (zetetic and poristic) is Analytic, the proof pro-ceeding by logical steps from something assumed as given to some given truth. They differ, however, in this: Zetetic leads the investigation to an equality (an equality, that is, of a given quantity with the sought quantity), whereas Poristic brings it to an identity, or to a given quantity itself, as one can see in examples of each kind. Hence there arises a further difference between them: in Poristic, since its progress is brought to an end by identity or by a given quantity, there is no need of a further process of resolution (as there is in Zetetic) for the final verification of the proposition.

DEFINITION 9

From the definition of Exegesis already given, it is evident that it needs to be twofold, in accordance with the twofold nature of Logistic—numerical and spe-cious. Although these deal with the same subject-matter, and have the same end in view—the resolution of equalities which have been already established—in practi-ce, however, and in mode of operation, they fall into entirely different genera. For, specious Exegesis takes the equality first established by means of Zetetic; then, by a continuous chain of inferences, it reduces it to a symbol or formal expression which is suitably arranged for resolution; and finally, from this properly orde-red expression, by means of a precise and straightforward technique, it reveals the sought quantity in its essential form. Indeed, when this form of Exegesis is applied in equations which do not go beyond the quadratic order but remain within a plane locus (as the ancients say), it should be regarded as perfectly scientific, because of the precise uniformity and determined nature of the method. When certain mo- 3 dern writers on Algebra have tried, however, to advance the art to the solution of higher order equations (that is, cubic and biquadratic) which are beyond the range of the previous method, they have left the method to us in a mutilated and incomplete state, because of ineradicable faults in their fundamental principles; as a consequence, a large proportion of the equations which fall within these higher categories have been regarded as insoluble.

DEFINITION 10

And so that other Exegesis—*numerical Exegesis*—was devised, which extends to the resolutions of all equations of all orders. It employs a general and infallible method: proceeding from any equation whatsoever which has been established by the resolving process of Zetetic and then returned to numerical form, the sought quantity is, by this method, revealed numerically by a secondary application of a different sort of resolution. This Exegesis has an art all of its own, provided with its own rules and instructions for practical application; they are given in the present treatise, which is wholly concerned with Exegesis.

DEFINITION 11

It seemed appropriate to call the numerical resolution of Exegesis *secondary*; secondary, that is, to the antecedent Zetetic resolution, which one secures first. In this way, by adding a qualification to the name that they share, it will be clear that both are a form of analysis or resolution, even if they are of different genera: zetetic is logical or discursive [resolution], while Exegesis, on the other hand, is instrumental. For, numerical Exegesis is precisely this: an elevation (so to speak) of the ancient practice of Arithmetic (which, until now, has gone no further in its usual application than the extraction of roots of only the simplest powers) to a new generality of method, not even attempted by the ancients.

DEFINITION 12

The word *root* is used in two senses in what follows. When establishing equalities through Zetetic, that which was assumed or taken as given at the beginning of the reasoning is, in fact, constantly sought; thus, in equations that have already been established, it can be called the sought or posited root—for as long, that is, as it remains concealed beneath the veils of the equation either in the form of a power or simply as itself. In resolving equations, on the other hand, or dealing with equations which have been already resolved—when it has been revealed by specious exegesis from an equation already established, by analytic reasoning and in its algebraic form; or when it has been elicited by numerical exegesis from the given homogeneous term of the equation, by analytic operation and in explicit numerical form—one may call it the revealed or extracted root (the variation in name corresponding to the variation in the type of resolution); this has also commonly come to be called the *value* of the sought root.

And so although in Vite's Exegesis, where problems about quadratics and cubes and higher powers are clearly dealt with, one inevitably had to inquire about the root of the power, in this work—which deals not with the powers, but with the equations themselves taken subjectively and in an integral form, in which the root is no more [the root] of the highest power than it is of the lower degrees—it seemed more in keeping with the stated intentions if the inquiry was directed to the explicatory root of the equation, or at least to the value of the sought root. Since, however, the explicatory root of the equation also generates the power, it is obvious that, to those who understand the matter itself, it will make no difference at all whether it is given this name or that.

DEFINITION 13

The definition given above of an equation should be taken as applying to one that has properly been established by Zetetic. Since such an equation is commonly **4** derived from the terms of problems, or of questions put forward in whatever other way, it may be called a *common* or *adventitious* equation; this is so that its name may distinguish it from the other, entirely different kind of equation, which will be called 'canonical'—and about which more will be said presently.

DEFINITION 14

There is also another kind of equations: those which, although not canonical, will be called *originals of canonical equations* in what follows, because canonical equations are derived from them, as from their originals. Equations of this kind are formed immediately, and without any discursive reasoning, from binomial roots by genesis or multiplication. In these cases, whatever is formed by multiplying the roots, is then set equal to those same roots (as they were before they were multiplied and only arranged in the proper way for multiplication), as one can see in these examples here.

$$\left.\begin{array}{l} a + b \\ a - c \end{array}\right| = \begin{array}{l} aa + ba \\ - ca - bc \end{array}$$

$$\left.\begin{array}{l} a + b \\ a + c \\ a - d \end{array}\right| = \begin{array}{l} aaa + baa + bca \\ + caa - bda \\ - daa - cda - bcd \end{array}$$

$$\left.\begin{array}{l} a + b \\ a + c \\ a + d \\ a - f \end{array}\right| = \begin{array}{l} aaaa + baaa + bcaa \\ + caaa + bdaa \\ + daaa + cdaa + bcda \\ - faaa - bfaa - bcfa \\ - cfaa - bdfa \\ - dfaa - cdfa - bcdf \end{array}$$

The form of these equations does not agree at all with the definition of a proper equation. But the two sorts of canonical equation which are derived from these, namely the primary and secondary, both conform to the definition, and are properly applicable in practice.

DEFINITION 15

The *primary* sort of *canonical equations* is of those which are established by derivation from original equations. The formal arrangement of the roots (the first part of an original equation) is suppressed; then, from the homogeneous terms of the second part, a given homogeneous term is transferred to the side opposite to the rest, by changing its sign—and thus a primary canonical equation is formed. The mode of derivation is set out in Section 2, but here are examples of the form.

$$aa + ba$$
$$- ca = +bc$$

$$aaa + baa + bca$$
$$+ caa - bda$$
$$- daa - cda = +bcd$$

$$aaaa + baaa + bcaa$$
$$+ caaa + bdaa$$
$$+ daaa + cdaa + bcda$$
$$- faaa - bfaa - bcfa$$
$$- cfaa - bdfa$$
$$- dfaa - cdfa = +bcdf$$

DEFINITION 16

5 The *secondary* sort of *canonical equations* is of those which are established by reduction from primary canonical equations. By removing any one of the incidental degrees it becomes a secondary or reduced one. This kind of reduction takes various forms, and they are discussed in the third section of this treatise; but some examples of this type follow:

$$aa = +bb$$

$$aaa - bba$$
$$- bca$$
$$- cca = + bbc$$
$$+ bcc$$

$$aaaa - bbaa - bbca$$
$$- bcaa - bcca$$
$$- ccaa - bbda$$
$$- bdaa - bdda$$
$$- cdaa - ccda$$
$$- ddaa - cdda$$
$$- 2.bcda = + bbcd$$
$$+ bccd$$
$$+ ccdd$$

DEFINITION 17

These two sorts of equations are regarded as canonical because, when they are applied as canons or yardsticks, the number of roots in common equations is determined (which can be seen in the Fifth Section); hence the name 'canonical' should be given to them not from the form in which they are established but from their use as instruments in this way.

DEFINITION 18

An equation is called reciprocal when a given homogeneous term is equated with one made from coefficients: and, reciprocally, when the [highest] power is equated with a term made from incidental [i.e., lower] degrees. As, for example, $aaa - caa + bba = +bbc$. For bbc is equated with $\dfrac{bb}{c}\,\Big|$ and aaa is equated with $\dfrac{aa}{a}\,\Big|$.

Division of this Analytic Treatise   6

In two parts of which

the first is preparatory to the exegesis, and is divided into 6 sections, namely...

the first: the forms of Specious Arithmetic are set out, together with practical examples.

the second: the derivation of primary canonical equations from their originals is demonstrated. Prefaced to this is a systematic description of the generation of the originals from binomial roots.

the third: this treats the reduction of secondary canonical equations from primary ones, by the removal of an incidental degree, while the substituted root remains unchanged.

the fourth: the explicatory roots of canonical equations (both primary and reduced) are indicated.

the fifth: the number of roots of common equations is determined by means of equipollent canonicals.

the sixth: this treats the reduction of common equations by the exclusion of an incidental degree and by the change of the substituted root.

the second contains the actual practice of numerical exegesis, explained through rules and examples; inasmuch as this is the principal skill of this art, anything treated in the previous part must be understood to be subordinate to this.

The Forms of the Four Operations of Specious Arithmetic[2] Illustrated by Example[3]

Examples of Addition

To be added	a	To be added	aa	To be added	aaa
	b		bc		bcc
Total	$a+b$	Total	$aa+bc$	Total	$aaa+bcc$
To be added	$a+b$	To be added		$a+b$
	$c+d$			$c-d$
Total	$a+b+c+d$	Total		$a+b+c-d$ [4]
To be added	$a+b$	To be added		$a+b$
	$-d$			$-b$
Total	$a+b-d$	Total		a
To be added	$a+b$	To be added		$aa+cc$
	$c+b$			$aa+cc$
Total	$a+c+2b$	Total		$2aa+2cc$ [5]
To be added	$aaa+cdf-ddd$	To be added		$b+7a$
	$aaa+bdd+ddd$			$+9a$
Total	$2aaa+cdf+bdd$	Total		$b+16a$
To be added	$b+7a$	To be added	$b+9a$	To be added	$b-9a$
	$-9a$		$b-7a$		$b+7a$
Total	$b-2a$	Total	$2b+2a$	Total	$2b-2a$

Examples of Subtraction

Given	a	Given	aa	Given	aaa
To be subtracted	b	To be subtracted	bc	To be subtracted	bcc
The remainder	$a-b$	The remainder	$aa-bc$	The remainder	$aaa-bcc$

				8
Given	$a+b$	Given	$a+b$	
To be subtracted	$a+d$	To be subtracted	$c-d$	
The remainder	$b-d$	The remainder	$a+b-c+d$	
Given	$a+b$	Given	$a+b$	
To be subtracted	$-d$	To be subtracted	$-b$	
The remainder	$a+b+d$	The remainder	$a+2b$	
Given	$a+b$	Given	$aa+cc$	
To be subtracted	$c+b$	To be subtracted	$aa+cc$	
The remainder	$a-c$	The remainder	0	
Given	$aaa+cdf-ddd$			
To be subtracted	$aaa+bdd+ddd$			
The remainder	$cdf-bdd-2ddd$			
Given	$b+7a$	Given	$b+7a$	
To be subtracted	$+9a$	To be subtracted	$-9a$	
The remainder	$b-2a$	The remainder	$b+16a$	
Given	$b+9a$	Given	$b-9a$	
To be subtracted	$b-7a$	To be subtracted	$b+7a$	
The remainder	$+16a$	The remainder	$-16a$	

Examples of Multiplication[6]

To be multiplied	a	To be multiplied	bc	
	b		d	
Product	ab	Product	bcd	
To be multiplied	aa	To be multiplied	bbb	
	bb		bb	
Product	$aabb$	Product	$bbbbb$	
To be multiplied	$bbcc$	To be multiplied	$b+a$	
	dd		$b+a$	
Product	$bbccdd$		$bb+ba$	
			$+ba+aa$	
		Product	$bb+2ba+aa$	
To be multiplied	$b-a$	To be multiplied	$b+a$	9
	$b-a$		$b-a$	
	$bb-ba$		$bb+ba$	
	$-ba+aa$		$-ba-aa$	
Product	$bb-2ba+aa$	Product	$bb-aa$	
To be multiplied	$b+c+d$	To be multiplied	$b+c-d$	
	a		$b-c+d$	
Product	$ba+ca+da$		$bb+bc-bd$	
			$-bc-cc+dc$	
			$+bd+dc-dd$	
		Product	$bb-cc+2cd-dd$	

Examples of Division or of Application[7]

Dividend	$bbcc$	Dividend	$bcdc$
Divisor	cc	Divisor	bdc
Quotient	bb	Quotient	c
Dividend	$bcdf$	Dividend	$ba + ca + da$
Divisor	cf	Divisor	a
Quotient	bd	Quotient	$b + c + d$
Dividend	$ba + ca + da$	Dividend	$bb + 2ba + aa$
Divisor	$b + c + d$	Divisor	$b + a$
Quotient	a	Quotient	$b + a$
Dividend	$bb - aa$	Dividend	$bb - aa$
Divisor	$b - a$	Divisor	$b + a$
Quotient	$b + a$	Quotient	$b - a$

The three last examples are evident from a generation [sc. of the dividend by multiplication] which is already known.

Note

If the operation of division is being considered in the guise of application [of areas], we can employ the terms 'applied area', 'measure' and 'result' for the words Dividend, Divisor and Quotient[8], or words similar to these, which better fit the way the operation is being visualized.

Symbols of Comparison to be Used in What Follows   **10**

The symbol for equality is $=$ so that $a = b$ will signify that a equals b.
The symbol for 'greater than' is $>$ so that $a > b$ will signify that a is greater than b.
The symbol for 'less than' is $<$ so that $a < b$ will signify that a is less than b[9].

Reducible Fractions Equated to Their Reduced Forms

$$\frac{ba}{b} = a \qquad \frac{bca}{b} = ca \qquad \frac{bca}{c} = ba \qquad \frac{bcda}{ca} = bd$$

$$\frac{ba}{c} + d = \frac{ba}{c} + \frac{dc}{c} = \frac{ba + dc}{c} \qquad \frac{ae}{b} + d = \frac{ae + db}{b}$$

$$\frac{ac}{b} + \frac{dd}{g} = \frac{acg}{bg} + \frac{bdd}{bg} = \frac{acg + bdd}{bg}$$

$$\frac{ac}{b} - d = \frac{ac}{b} - \frac{db}{b} = \frac{ac - db}{b}$$

$$\frac{ac}{b} - \frac{dd}{g} = \frac{acg}{bg} - \frac{ddb}{bg} = \frac{acg - ddb}{bg}$$

$$\left.\frac{\frac{ac}{b}}{b}\right| = \frac{acb}{b} = ac \qquad \left.\frac{\frac{ac}{b}}{d}\right| = \frac{acd}{b} \qquad \left.\frac{\frac{ac}{b}}{\frac{dd}{g}}\right| = \frac{acdd}{bg}$$

$$\frac{\frac{aaa}{b}}{d} = \frac{aaa}{bd} \qquad \frac{bg}{\frac{ac}{d}} = \frac{bgd}{ac} \qquad \frac{\frac{bbb}{c}}{\frac{aaa}{dg}} = \frac{bbbdg}{caaa}$$

Reductions of Irregular Equations to an appropriate Form, Illustrated by Example   11

By Antithesis[10] or Transposition of the Elements, which comes about by uniform Addition

Let $aa - dc = gg$ be the equation to be reduced.
Let $+dc$ be added to both sides.
Hence it becomes $aa = gg + dc$, the reduced equation.

Likewise let $aa - dc = gg - ba$ be the equation to be reduced.
Let $+ba + dc$ be added to both sides.
Thence it becomes $aa + ba = gg + dc$, the reduced equation.

By uniform divison, by which a homogeneous term in given quantities is released out of a term involving some power [of the unknown]; this is the Hypobibasmus *of Viète*[11]

Let $aaa + baa = dca$ be the equation to be reduced.
Therefore $\frac{aaa}{a} + \frac{baa}{a} = \frac{dca}{a}$.
Therefore $aa + ba = dc$ is the reduced equation.

By uniform division, by which a power [of the unknown] is released out of a term involving some power [of the known quantities]; this is the Parabolismus *of Viète*[12]

Let $baa + dca = gcd$ be the equation to be reduced.
Therefore $\frac{baa}{b} + \frac{dca}{b} = \frac{gcd}{b}$.
Therefore $aa + \frac{dca}{b} = \frac{gcd}{b}$ is the reduced equation.

Or let $baa + dba = cbd$ be the equation to be reduced.
Therefore $\frac{baa}{b} + \frac{dba}{b} = \frac{cbd}{b}$.
Therefore $aa + da = cd$ is the reduced equation.

Section Two[1]

Derivation or Deduction of Canonical[2] Equations from their originals[3]

In preface, an ordered description of the original equations themselves as they arise from binomial roots through multiplying out

Quadratics

1. $\begin{array}{|c}a - b \\ a - c\end{array} = \begin{array}{l} aa - ba \\ \quad - ca + bc \end{array}$ [4]

2. $\begin{array}{|c}a - b \\ a + c\end{array} = \begin{array}{l} aa - ba \\ \quad + ca - bc \end{array}$

3. $\begin{array}{|c}a + b \\ a + c\end{array} = \begin{array}{l} aa + ba \\ \quad + ca + bc \end{array}$ [5]

Cubics[6]

1. $\begin{array}{|c}a - b \\ a - c \\ a - d\end{array} = \begin{array}{l} aaa - baa + bca \\ \quad - caa + bda \\ \quad - daa + cda - bcd \end{array}$

2. $\begin{array}{|c}a - b \\ a - c \\ a + d\end{array} = \begin{array}{l} aaa - baa + bca \\ \quad - caa - bda \\ \quad + daa - cda + bcd \end{array}$

3. $\begin{array}{|c}a + b \\ a + c \\ a - d\end{array} = \begin{array}{l} aaa + baa + bca \\ \quad + caa - bda \\ \quad - daa - cda - bcd \end{array}$

4. $\begin{array}{c} a+b \\ a+c \\ \underline{a+d} \end{array}\bigg| = \begin{array}{l} aaa+baa+bca \\ +caa+bda \\ +daa+cda+bcd \end{array}$

Reciprocal forms of Cubics

5. $\begin{array}{c} aa-bc \\ \underline{a-d} \end{array}\bigg| = aaa-daa-bca+bcd$

6. $\begin{array}{c} aa+bc \\ \underline{a-d} \end{array}\bigg| = aaa-daa+bca-bcd$

13   7. $\begin{array}{c} aa-bc \\ \underline{a+d} \end{array}\bigg| = aaa+daa-bca-bcd$

8. $\begin{array}{c} aa+bc \\ \underline{a+d} \end{array}\bigg| = aaa+daa+bca+bcd$ [7]

Three Cubic varieties of canonical equations, originally constructed out of roots set equal [to a quantity]

Let $r - a = q$.

9. Then $\begin{array}{c} r-a \\ r-a \\ r-a \end{array}\bigg| = rrr - 3rra + 3raa - aaa = +qqq.$ [8]

Let $r + a = q$.

10. Then $\begin{array}{c} r+a \\ r+a \\ r+a \end{array}\bigg| = rrr + 3rra + 3raa + aaa = +qqq.$

Let $a - r = q$.

11. Then $\begin{array}{c} a-r \\ a-r \\ a-r \end{array}\bigg| = aaa - 3raa + 3rra - rrr = +qqq.$

Biquadratics

1. $\begin{array}{c} a-b \\ a-c \\ a-d \\ \underline{a-f} \end{array}\bigg| = \begin{array}{l} aaaa-baaa+bcaa \\ -caaa+bdaa \\ -daaa+cdaa-bcda \\ -faaa+bfaa-bcfa \\ +cfaa-bdfa \\ +dfaa-cdfa+bcdf \end{array}$

2. $\left.\begin{array}{l} a-b \\ a-c \\ a-d \\ a+f \end{array}\right|$ $=$ $aaaa - baaa + bcaa$
$\qquad - caaa + bdaa$
$\qquad - daaa + cdaa - bcda$
$\qquad + faaa - bfaa + bcfa$
$\qquad\qquad - cfaa + bdfa$
$\qquad\qquad - dfaa + cdfa - bcdf$

3. $\left.\begin{array}{l} a+b \\ a+c \\ a+d \\ a-f \end{array}\right|$ $=$ $aaaa + baaa + bcaa$
$\qquad + caaa + bdaa$
$\qquad + daaa + cdaa + bcda$
$\qquad - faaa - bfaa - bcfa$
$\qquad\qquad - cfaa - bdfa$
$\qquad\qquad - dfaa - cdfa - bcdf$

14

4. $\left.\begin{array}{l} a-b \\ a-c \\ a+d \\ a+f \end{array}\right|$ $=$ $aaaa - baaa + bcaa$
$\qquad - caaa - bdaa$
$\qquad + daaa - cdaa + bcda$
$\qquad + faaa - bfaa + bcfa$
$\qquad\qquad - cfaa - bdfa$
$\qquad\qquad + dfaa - cdfa + bcdf$

5. $\left.\begin{array}{l} a+b \\ a+c \\ a+d \\ a+f \end{array}\right|$ $=$ $aaaa + baaa + bcaa$
$\qquad + caaa + bdaa$
$\qquad + daaa + cdaa + bcda$
$\qquad + faaa + bfaa + bcfa$
$\qquad\qquad + cfaa + bdfa$
$\qquad\qquad + dfaa + cdfa + bcdf$

Reciprocal forms of biquadratics

6. $\left.\begin{array}{l} aaa - cdf \\ a - b \end{array}\right|$ $= aaaa - baaa - cdfa + bcdf$

7. $\left.\begin{array}{l} aaa - cdf \\ a + b \end{array}\right|$ $= aaaa + baaa - cdfa - bcdf$

8. $\left.\begin{array}{l} aaa + cdf \\ a - b \end{array}\right|$ $= aaaa - baaa + cdfa - bcdf$

9. $\left.\begin{array}{l} aaa + cdf \\ a + b \end{array}\right|$ $= aaaa + baaa + cdfa + bcdf$

Some other biquadratic varieties of original equations

10. $\left.\begin{array}{l} b - a \\ c - a \\ df - aa \end{array}\right|$ $=$ $bcdf - bdfa + dfaa + baaa$
$\qquad - cdfa - bcaa + caaa - aaaa$

11. $\begin{vmatrix} b-a \\ c-a \\ df+aa \end{vmatrix} = \begin{aligned} &bcdf - bdfa + dfaa - baaa \\ &\quad - cdfa + bcaa - caaa + aaaa \end{aligned}$

15 12. $\begin{vmatrix} b+a \\ c+a \\ df-aa \end{vmatrix} = \begin{aligned} &bcdf + bdfa + dfaa - baaa \\ &\quad + cdfa - bcaa - caaa - aaaa \end{aligned}$

13. $\begin{vmatrix} b-a \\ c+a \\ df+aa \end{vmatrix} = \begin{aligned} &bcdf + bdfa - dfaa + baaa \\ &\quad - cdfa + bcaa - caaa - aaaa \end{aligned}$

14. $\begin{vmatrix} b+a \\ c-a \\ df-aa \end{vmatrix} = \begin{aligned} &bcdf - bdfa - dfaa + baaa \\ &\quad + cdfa - bcaa - caaa + aaaa \end{aligned}$

15. $\begin{vmatrix} b+a \\ c+a \\ df+aa \end{vmatrix} = \begin{aligned} &bcdf + bdfa + dfaa + baaa \\ &\quad + cdfa + bcaa + caaa + aaaa \end{aligned}$

16. $\begin{vmatrix} bc-aa \\ df-aa \end{vmatrix} = \begin{aligned} &bcdf - dfaa \\ &\quad - bcaa + aaaa \end{aligned}$

17. $\begin{vmatrix} bc-aa \\ df+aa \end{vmatrix} = \begin{aligned} &bcdf - dfaa \\ &\quad + bcaa - aaaa \end{aligned}$

18. $\begin{vmatrix} bc+aa \\ df+aa \end{vmatrix} = \begin{aligned} &bcdf + dfaa \\ &\quad + bcaa + aaaa \end{aligned}$

16 Derivation of Canonical Equations of the Quadratic Order

PROPOSITION 1

The canonical equation $\begin{aligned} aa - ba \\ + ca \end{aligned} = +bc$ is derived from the original equation $\begin{vmatrix} a-b \\ a+c \end{vmatrix} = \begin{aligned} aa - ba \\ + ca - bc \end{aligned}$ by putting b equal to a.

For, if we put $a = b$, then it will be true that $a - b = 0$. [9]

Consequently, putting $a = b$, it will be true that $\begin{vmatrix} a-b \\ a+c \end{vmatrix} = 0$.

Moreover, it is true by multiplying out that $\begin{vmatrix} a-b \\ a+c \end{vmatrix} = \begin{aligned} aa - ba \\ + ca - bc \end{aligned}$ which is the original equation designated in the proposition.

Therefore $\begin{aligned} aa - ba \\ + ca - bc \end{aligned} = 0$.

Therefore $\begin{aligned} aa - ba \\ + ca \end{aligned} = +bc$ which is the proposed equation.

Thus the proposed canonical equation is derived from the designated original equation, by putting b equal to a, as was stated in the proposition.[10]

PROPOSITION 2

The canonical equation $\begin{matrix} aa - ba \\ -ca \end{matrix} = -bc$ is derived from the original equation

$\begin{vmatrix} a - b \\ a - c \end{vmatrix} = \begin{matrix} aa - ba \\ -ca + bc \end{matrix}$ by putting b or c equal to a[11].

For, if we put $a = b$, it will be true that $a - b = 0$;

or if $a = c$, then it will be true that $a - c = 0$.

Consequently, putting $a = b$ or c, it will be true that $\begin{vmatrix} a - b \\ a - c \end{vmatrix} = 0$.

Moreover, it is true by multiplying out that $\begin{vmatrix} a - b \\ a - c \end{vmatrix} = \begin{matrix} aa - ba \\ -ca + bc \end{matrix}$ which is

the original equation designated in the proposition.

Therefore $\begin{matrix} aa - ba \\ -ca + bc \end{matrix} = 0$.

Therefore $\begin{matrix} aa - ba \\ -ca \end{matrix} = -bc$ which is the proposed equation.[12]

Thus the proposed canonical equation is derived from the designated original equation by putting b or c equal to a, as was stated in the proposition.

17

The Derivation of Canonical Equations of the Cubic Order

PROPOSITION 3

The canonical equation $\begin{matrix} aaa - baa - bca \\ +caa - bda \\ +daa + cda = +bcd \end{matrix}$ is derived from the

original equation $\begin{vmatrix} a - b \\ a + c \\ a + d \end{vmatrix} = \begin{matrix} aaa - baa - bca \\ +caa - bda \\ +daa + cda - bcd \end{matrix}$ by putting b equal to a.

For, if we put $a = b$ it will be true that $a - b = 0$.

Consequently, putting $a = b$, it is true that $\begin{vmatrix} a - b \\ a + c \\ a + d \end{vmatrix} = 0$.

Moreover it is true by multiplying out that $\begin{vmatrix} a - b \\ a + c \\ a + d \end{vmatrix} = \begin{matrix} aaa - baa - bca \\ +caa - bda \\ +daa + cda - bcd \end{matrix}$

which is the original equation designated in the proposition.

Therefore $\begin{matrix} aaa - baa - bca \\ +caa - bda \\ +daa + cda - bcd = 0 \end{matrix}$.

Therefore $\begin{matrix} aaa - baa - bca \\ +caa - bda \\ +daa + cda = +bcd \end{matrix}$ which is the proposed equation.

Thus the proposed canonical equation is derived from the designated original equation, by putting b equal to a, as was stated in the proposition.

PROPOSITION 4

The canonical equation
$$\begin{array}{l} aaa - baa + bca \\ -\,caa - bda \\ +\,daa - cda = -bcd \end{array}$$
is derived from the

original equation
$$\begin{array}{c|l} a - b \\ a - c \\ a + d \end{array} \;\; \begin{array}{l} = aaa - baa + bca \\ -\,caa - bda \\ +\,daa - cda + bcd \end{array}$$
by putting b or c equal to the root a.

18 For, if we put $b = a$ it will be true that $a - b = 0$;
or if $c = a$, it will be true that $a - c = 0$.

Consequently, putting b or $c = a$, it will be true that
$$\begin{array}{c|} a - b \\ a - c \\ a + d \end{array} = 0.$$

Moreover it is true by multiplying out that
$$\begin{array}{c|l} a - b \\ a - c \\ a + d \end{array} \;\; \begin{array}{l} = aaa - baa + bca \\ -\,caa - bda \\ +\,daa - cda + bcd \end{array}$$
which is the original equation designated in the proposition.

Therefore
$$\begin{array}{l} aaa - baa + bca \\ -\,caa - bda \\ +\,daa - cda + bcd = 0. \end{array}$$

Therefore
$$\begin{array}{l} aaa - baa + bca \\ -\,caa - bda \\ +\,daa - cda = -bcd \end{array}$$
which is the proposed equation.

Thus the proposed canonical equation is derived from the designated original equation, by putting b or c equal to a, as was stated in the proposition.

PROPOSITION 5

The canonical equation
$$\begin{array}{l} aaa - baa + bca \\ -\,caa + bda \\ -\,daa + cda = +bcd \end{array}$$
is derived from the

original equation
$$\begin{array}{c|l} a - b \\ a - c \\ a - d \end{array} \;\; \begin{array}{l} = aaa - baa + bca \\ -\,caa + bda \\ -\,daa + cda - bcd \end{array}$$
by putting b or c or d equal to a.

For, if we put $b = a$ it will be true that $a - b = 0$;

or if $c = a$, it will be true that $a - c = 0$;

or if $d = a$, it will be true that $a - d = 0$.

Consequently, putting b or c or $d = a$, it will be true that
$$\begin{array}{c|} a - b \\ a - c \\ a - d \end{array} = 0.$$

Moreover it is true by multiplying out that
$$\begin{array}{c|l} a - b \\ a - c \\ a - d \end{array} \;\; \begin{array}{l} = aaa - baa + bca \\ -\,caa + bda \\ -\,daa + cda - bcd \end{array}$$
which is the original equation designated in the proposition.

Therefore
$$\begin{array}{l} aaa - baa + bca \\ -\,caa + bda \\ -\,daa + cda - bcd = 0 \end{array}$$

Therefore $\begin{array}{l} aaa - baa + bca \\ \quad - caa + bda \\ \quad - daa + cda = +bcd \end{array}$ which is the proposed canonical equation.   **19**

Thus the proposed canonical equation is derived from the designated original equation, by putting b or c or d equal to a, as was stated in the proposition.

Derivation of Reciprocal forms of the Cubic Order

PROPOSITION 6

The reciprocal equation $aaa - baa + cda = +bcd$ is derived from the original

equation $\left.\dfrac{a - b}{aa + cd}\right| = aaa - baa + cda - bcd$ by putting b equal to a.[13]

For, if we put $b = a$ it will be true that $a - b = 0$.

Consequently, putting $b = a$ it is true that $\left.\dfrac{a - b}{aa + cd}\right| = 0.$

Moreover it is true by multiplying out that $\left.\dfrac{a - b}{aa + cd}\right| = aaa - baa + cda - bcd$

which is the original equation designated in the proposition.

Therefore $aaa - baa + cda - bcd = 0.$

Therefore $aaa - baa + cda = +bcd$ which is the proposed reciprocal equation.

Thus the proposed reciprocal equation is derived from the designated original equation, by putting b equal to a, as was stated in the proposition.

PROPOSITION 7

The reciprocal equation $aaa + baa - cda = +bcd$ is derived from the original

equation $\left.\dfrac{a + b}{aa - cd}\right| = aaa + baa - cda - bda$ by putting $cd = aa$.

For, if we put $cd = aa$ it will be true that $aa - cd = 0$.

Consequently, putting $cd = aa$ it is true that $\left.\dfrac{a + b}{aa - cd}\right| = 0.$

Moreover it is true by multiplying out that $\left.\dfrac{a + b}{aa - cd}\right| = aaa + baa - cda - bcd$

which is the original equation designated in the proposition.

Therefore $aaa + baa - cda - bcd = 0.$ **20**

Therefore $aaa + baa - cda = +bcd$ which is the proposed equation.

Thus the proposed reciprocal equation is derived from the designated original equation, by putting $cd = aa$, as was stated in the proposition.

PROPOSITION 8

The reciprocal equation $aaa - baa - cda = -bcd$ is derived from the original

equation $\left.\dfrac{a - b}{aa - cd}\right| = aaa - baa - cda + bcd$ by putting $b = a$ or $cd = aa$.

For, if we put $b = a$ it will be true that $a - b = 0$;

or if $cd = aa$, it will be true that $aa - cd = 0.$

Consequently putting $b = a$ or $cd = aa$, it is true that $\begin{vmatrix} a - b \\ aa - cd \end{vmatrix} = 0$.

Moreover it is true by multiplying out that $\begin{vmatrix} a - b \\ aa - cd \end{vmatrix} = aaa - baa - cda + bcd$

which is the original equation designated in the proposition.

Therefore $aaa - baa - cda + bcd = 0$.

Therefore $aaa - baa - cda = -bcd$ which is the proposed reciprocal equation.

Thus the proposed reciprocal equation is derived from the designated original equation, by putting $b = a$ or $cd = aa$, as was stated in the proposition.

Derivation of Canonical Biquadratics[14]

PROPOSITION 9

The canonical equation
$$\begin{aligned}
&aaaa - baaa\ - bcaa \\
&\quad + caaa\ - bdaa \\
&\quad + daaa - bfaa - bcda \\
&\quad + faaa + cdaa - bcfa \\
&\quad\quad + cfaa - bdfa \\
&\quad\quad + dfaa + cdfa = +bcdf
\end{aligned}$$

21 is derived from the original equation

$$\begin{vmatrix} a - b \\ a + c \\ a + d \\ a + f \end{vmatrix} = \begin{aligned}
&aaaa - baaa\ - bcaa \\
&\quad + caaa\ - bdaa \\
&\quad + daaa - bfaa - bcda \\
&\quad + faaa + cdaa - bcfa \\
&\quad\quad + cfaa - bdfa \\
&\quad\quad + dfaa + cdfa - bcdf
\end{aligned} \qquad \text{by putting } b = a.$$

For, if we put $b = a$ it will be true that $a - b = 0$.

Consequently, putting $b = a$ it is true that $\begin{vmatrix} a - b \\ a + c \\ a + d \\ a + f \end{vmatrix} = 0$.

Moreover it is true by multiplying out that

$$\begin{vmatrix} a - b \\ a + c \\ a + d \\ a + f \end{vmatrix} = \begin{aligned}
&aaaa - baaa\ - bcaa \\
&\quad + caaa\ - bdaa \\
&\quad + daaa - bfaa - bcda \\
&\quad + faaa + cdaa - bcfa \\
&\quad\quad + cfaa - bdfa \\
&\quad\quad + dfaa + cdfa - bcdf
\end{aligned} \qquad \text{which is the original}$$

equation designated in the proposition.

Therefore
$$\begin{aligned}
&aaaa - baaa\ - bcaa \\
&\quad + caaa\ - bdaa \\
&\quad + daaa - bfaa - bcda \\
&\quad + faaa + cdaa - bcfa \\
&\quad\quad + cfaa - bdfa \\
&\quad\quad + dfaa + cdfa - bcdf = 0.
\end{aligned}$$

Therefore
$$\begin{aligned}
&aaaa - baaa\ - bcaa \\
&\quad + caaa\ - bdaa \\
&\quad + daaa - bfaa - bcda \\
&\quad + faaa + cdaa - bcfa \\
&\quad\quad + cfaa - bdfa \\
&\quad\quad + dfaa + cdfa = +bcdf
\end{aligned} \qquad \text{which is the proposed}$$

equation.

Thus the proposed canonical equation is derived from the designated original equation, by putting $b = a$, as was stated in the proposition.

PROPOSITION 10

The canonical equation

$$
\begin{aligned}
aaaa &- baaa + bcaa \\
&- caaa - bdaa \\
&+ daaa - bfaa + bcda \\
&+ faaa - cdaa + bcfa \\
&- cfaa - bdfa \\
&+ dfaa - cdfa = -bcdf.
\end{aligned}
$$

is derived

from the original equation

$$
\begin{vmatrix} a - b \\ a - c \\ a + d \\ a + f \end{vmatrix} =
\begin{aligned}
aaaa &- baaa + bcaa \\
&- caaa - bdaa \\
&+ daaa - bfaa + bcda \\
&+ faaa - cdaa + bcfa \\
&- cfaa - bdfa \\
&+ dfaa - cdfa + bcdf
\end{aligned}
$$

by putting $b = a$ or $c = a$.

For if we put $b = a$ it will be true that $a - b = 0$.

Or $c = a$ it will be true that $a - c = 0$.

Consequently putting b or $c = a$ it is true that $\begin{vmatrix} a - b \\ a - c \\ a + d \\ a + f \end{vmatrix} = 0.$

Moreover it is true by multiplying out that

$$
\begin{vmatrix} a - b \\ a - c \\ a + d \\ a + f \end{vmatrix} =
\begin{aligned}
aaaa &- baaa + bcaa \\
&- caaa - bdaa \\
&+ daaa - bfaa + bcda \\
&+ faaa - cdaa + bcfa \\
&- cfaa - bdfa \\
&+ dfaa - cdfa + bcdf
\end{aligned}
$$

which is the original equation designated in the proposition.

Therefore

$$
\begin{aligned}
aaaa &- baaa + bcaa \\
&- caaa - bdaa \\
&+ daaa - bfaa + bcda \\
&+ faaa - cdaa + bcfa \\
&- cfaa - bdfa \\
&+ dfaa - cdfa + bcdf = 0.
\end{aligned}
$$

Therefore

$$
\begin{aligned}
aaaa &- baaa + bcaa \\
&- caaa - bdaa \\
&+ daaa - bfaa + bcda \\
&+ faaa - cfaa + bcfa \\
&- cdaa - bdfa \\
&+ dfaa - cdfa = -bcdf.
\end{aligned}
$$

which is the proposed

equation.

Thus the proposed canonical equation is derived from the designated original equation, by putting b or $c = a$, as was stated in the proposition.

PROPOSITION 11

23 The canonical equation

$$
\begin{aligned}
aaaa &- baaa + bcaa \\
&- caaa + bdaa \\
&- daaa + cdaa - bcda \\
&+ faaa - bfaa + bcfa \\
&\quad\quad\;\; - cfaa + bdfa \\
&\quad\quad\;\; - dfaa + cdfa = +bcdf
\end{aligned}
$$

is derived from the original equation

$$
\left.\begin{array}{l}
a - b \\
a - c \\
a - d \\
a + f
\end{array}\right| =
\begin{aligned}
aaaa &- baaa + bcaa \\
&- caaa + bdaa \\
&- daaa + cdaa - bcda \\
&+ faaa - bfaa + bcfa \\
&\quad\quad\;\; - cfaa + bdfa \\
&\quad\quad\;\; - dfaa + cdfa - bcdf
\end{aligned}
$$

by putting b or c or $d = a$.

For if we put $b = a$ it will be true that $a - b = 0$.

Or $c = a$ it will be true that $a - c = 0$.

Or $d = a$ it will be true that $a - d = 0$.

Consequently putting b or c or $d = a$ it is true that $\left.\begin{array}{l} a - b \\ a - c \\ a - d \\ a + f \end{array}\right| = 0.$

Moreover it is true by multiplying out that

$$
\left.\begin{array}{l}
a - b \\
a - c \\
a - d \\
a + f
\end{array}\right| =
\begin{aligned}
aaaa &- baaa + bcaa \\
&- caaa + bdaa \\
&- daaa + cdaa - bcda \\
&+ faaa - bfaa + bcfa \\
&\quad\quad\;\; - cfaa + bdfa \\
&\quad\quad\;\; - dfaa + cdfa - bcdf.
\end{aligned}
$$

which is the original equation designated in the proposition.

24 Therefore

$$
\begin{aligned}
aaaa &- baaa + bcaa \\
&- caaa + bdaa \\
&- daaa + cdaa - bcda \\
&+ faaa - bfaa + bcfa \\
&\quad\quad\;\; - cfaa + bdfa \\
&\quad\quad\;\; - dfaa + cdfa - bcdf = 0
\end{aligned}
$$

Therefore

$$
\begin{aligned}
aaaa &- baaa + bcaa \\
&- caaa + bdaa \\
&- daaa + cdaa - bcda \\
&+ faaa - bfaa + bcfa \\
&\quad\quad\;\; - cfaa + bdfa \\
&\quad\quad\;\; - dfaa + cdfa = +bcdf.
\end{aligned}
$$

which is the proposed

equation.

Thus the proposed canonical equation is derived from the designated original equation, by putting b or c or d $= a$, as was stated in the proposition.

PROPOSITION 12

The canonical equation

$$\begin{aligned}
aaaa &- baaa + bcaa\\
&- caaa + bdaa\\
&- daaa + bfaa - bcda\\
&- faaa + cdaa - bcfa\\
&+ cfaa - bdfa\\
&+ dfaa - cdfa = -bcdf
\end{aligned}$$

is derived from the original equation

$$\left.\begin{array}{c}
a - b\\
a - c\\
a - d\\
a - f
\end{array}\right| = \begin{aligned}
aaaa &- baaa + bcaa\\
&- caaa + bdaa\\
&- daaa + bfaa - bcda\\
&- faaa + cdaa - bcfa\\
&+ cfaa - bdfa\\
&+ dfaa - cdfa + bcdf
\end{aligned}$$

by putting b or c or d or $f = a$.

For if we put $b = a$ it will be true that $a - b = 0$.
Or $c = a$ it will be true that $a - c = 0$.
Or $d = a$ it will be true that $a - d = 0$.
Or $f = a$ it will be true that $a - f = 0$.

Consequently putting b or c or d or $f = a$ it is true that $\left.\begin{array}{c} a - b\\ a - c\\ a - d\\ a - f \end{array}\right| = 0.$

Moreover it is true by multiplying out that   25

$$\left.\begin{array}{c}
a - b\\
a - c\\
a - d\\
a - f
\end{array}\right| = \begin{aligned}
aaaa &- baaa + bcaa\\
&- caaa + bdaa\\
&- daaa + bfaa - bcda\\
&- faaa + cdaa - bcfa\\
&+ cfaa - bdfa\\
&+ dfaa - cdfa + bcdf.
\end{aligned}$$

which is the original equation designated in the proposition.

Therefore

$$\begin{aligned}
aaaa &- baaa + bcaa\\
&- caaa + bfaa\\
&- daaa + bfaa - bcda\\
&- faaa + cdaa - bcfa\\
&+ cfaa - bdfa\\
&+ dfaa - cdfa + bcdf = 0.
\end{aligned}$$

Therefore

$$\begin{aligned}
aaaa &- baaa + bcaa\\
&- caaa + bdaa\\
&- daaa + bfaa - bcda\\
&- faaa + cdaa - bcfa\\
&+ cfaa - bdfa\\
&+ dfaa - cdfa = -bcdf.
\end{aligned}$$

which is the proposed ca-

nonical equation.

Thus the proposed canonical equation is derived from the designated original equation, by putting b or c or d or $f = a$, as was stated in the proposition.

The Derivation of Reciprocal Equations of the Biquadratic Order

PROPOSITION 13

The reciprocal equation $aaaa - baaa + cdfa = +bcdf$ is derived from the original equation $\overline{\dfrac{aaa + cdf}{a - b}} = aaaa - baaa + cdfa - bcdf$ by putting $b = a$.

For if we put $b = a$ it will be true that $a - b = 0$. Consequently putting $b = a$ it is true that $\overline{\dfrac{aaa + cdf}{a - b}} = 0$.

Moreover it is true by multiplying out that $\overline{\dfrac{aaa + cdf}{a - b}} = aaaa - baaa + cda +$ $cdfa - bcdf$ which is the original equation designated in the proposition. Therefore $aaaa - baaa + cdfa - bcdf = 0$. Therefore $aaaa - baaa + cdfa = +bcdf$ which is the proposed reciprocal equation.

Thus the proposed reciprocal equation is derived from the designated original equation, by putting $b = a$, as was stated in the proposition.

26
PROPOSITION 14

The reciprocal equation $aaaa + baaa - cdfa = +bcdf$ is derived from the original equation $\overline{\dfrac{aaa - cdf}{a + b}} = aaaa + baaa - cdfa - bcdf$ by putting $cdf = aaa$.

For if we put $cdf = aaa$. it will be true that $aaa - cdf = 0$.

Consequently putting $cdf = aaa$ it is true that $\overline{\dfrac{aaa - cdf}{a + b}} = 0$.

Moreover it is true by multiplying out that $\overline{\dfrac{aaa - cdf}{a + b}} = aaaa + baaa - cdfa -$ $bcdf$ which is the original equation designated in the proposition.

Therefore $aaaa + baaa - cdfa - bcdf = 0$.

Therefore $aaaa + baaa - cdfa = +bcdf$ which is the proposed reciprocal equation.

Thus the proposed reciprocal equation is derived from the designated original equation, by putting $cdf = aaa$, as was stated in the proposition.

PROPOSITION 15

The reciprocal equation $aaaa - baaa - cdfa = -bcdf$ is derived from the original equation $\overline{\dfrac{aaa - cdf}{a - b}} = aaaa - baaa - cdfa + bcdf$ by putting $cdf = aaa$ or $b = a$.

For if we put $b = a$ it will be true that $a - b = 0$. Or $cdf = aaa$ it will be true that $aaa - cdf = 0$.

Consequently putting $b = a$. Or $cdf = aaa$ it is true that $\overline{\dfrac{aaa - cdf}{a - b}} = 0$.

Moreover it is true by multiplying out that $\left| \dfrac{aaa - cdf}{a - b} \right| = aaaa - baaa - cdfa +$
$bcdf$ which is the original equation designated in the proposition.

Therefore $aaaa - baaa - cdfa + bcdf = 0$.

Therefore $aaaa - baaa - cdfa = -bcdf$ which is the proposed reciprocal equation.

Thus the proposed reciprocal equation is derived from the designated original equation, by putting $b = a$ or $cdf = aaa$, as was stated in the proposition.

Note[15]

The derivation of canonical equations of the biquadratic order from the last eight types of original equations (that is, types 10. 11. 12. 13. 14. 15. 16. 17.) is sufficiently clear from the example of the propositions above.

However, since the derivations of canonical equations from the following original equations: 3. of the Quadratics; 4. & 8. of the Cubics; 5. 9. & 18. of the Biquadratics, cannot be done without the substitution of negative roots, they are omitted insofar as they are useless.

Also, the three equations of the cubic order which are generated by setting the roots equal [to a quantity] can be included here as if they were derived canonical equations, thus (leaving out the formal arrangement of the roots and the symbol for multiplying out):[16]

Let $b - a = c$.

Then $aaa - 3baa + 3bba = -ccc + bbb$

Let $a + b = c$.

Then $aaa + 3baa + 3bba = +ccc - bbb$

Let $a - b = c$.

Then $aaa - 3baa + 3bba = +ccc + bbb$.

Summary of the Canonical Equations whose Derivations were demonstrated in this Second Section[17]

Quadratics

1. $aa - ba = +bc$
   $ + ca$

2. $aa - ba = -bc$
   $ - ca$

Cubics

1. $aaa - baa - bca$
   $ + caa - bda$
   $ + daa + cda = +bcd$

2. $aaa - baa + bca$
   $- caa - bda$
   $+ daa - cda = -bcd$

3. $aaa - baa - bca$
   $- caa - bda$
   $- daa - cda = +bcd$

4. $aaa - baa + cda = +bcd$

5. $aaa + baa - cda = +bcd$

6. $aaa - baa - cda = -bcd$

28 Biquadratics

1. $aaaa - baaa - bcaa$
   $+ caaa - bdaa$
   $+ daaa - bfaa - bcda$
   $+ faaa + cdaa - bcfa$
   $+ cfaa - bdfa$
   $+ dfaa + cdfa = +bcdf$

2. $aaaa - baaa + bcaa$
   $- caaa - bdaa$
   $+ daaa - bfaa + bcda$
   $+ faaa - cdaa + bcfa$
   $- cfaa - bdfa$
   $+ dfaa - cdfa = -bcdf$

3. $aaaa - baaa + bcaa$
   $- caaa + bdaa$
   $- daaa + cdaa - bcda$
   $+ faaa - bfaa + bcfa$
   $- cfaa + bdfa$
   $- dfaa + cdfa = +bcdf$

4. $aaaa - baaa + bcaa$
   $- caaa + bdaa$
   $- daaa + bfaa - bcda$
   $- faaa + cdaa - bcfa$
   $+ cfaa - bdfa$
   $+ dfaa - cdfa = -bcdf$

5. $aaaa - baaa + cdfa = +bcdf$

6. $aaaa + baaa - cdfa = +bcdf$

7. $aaaa - baaa - cdfa = -bcdf$

Section Three[1]

The Reduction of Secondary Canonical Equations from Primary [canonical equations] by removing a subsidiary degree, while leaving the posited root unchanged.

The Single Reduction of a Canonical Quadratic Equation

PROBLEM 1[2]

To reduce the binomial equation $\begin{array}{l} aa + ba \\ \quad + ca \end{array} = +bc$ to the mononomial $aa = bb$, that is, by removing the first degree a.

Let us put $b = c$.

And in the binomial equation which is to be reduced, let c be changed into b.

From which it follows that $\begin{array}{l} aa - ba \\ \quad + ba \end{array} = +bb$.

And so, removing the terms which are redundant because of their opposite signs, it becomes $aa = bb$, the mononomial equation which was sought.

Thus the required reduction of the proposed binomial equation to the mononomial which was sought is, in this way, accomplished.

Reductions of Canonical Equations of the Cubic Order

PROBLEM 2

To reduce the trinomial equation $\begin{array}{l} aaa - baa + bca \\ \quad - caa - bda \\ \quad + daa - cda \end{array} = -bcd$

to the binomial $\begin{array}{l} aaa - bba \\ \quad - bca \\ \quad - cca \\ \quad - bcc, \end{array} = -bbc$ that is, by removing the second degree aa.

Let us put $b + c = d$.

And in the trinomial equation which is to be reduced, let d be changed into $b + c$.

30 From which it follows that

$$aaa - baa + bca$$
$$- caa - bba$$
$$+ baa - bca$$
$$+ caa - bca$$
$$- cca = - bbc$$
$$- bcc.$$

And so, removing the terms which are redundant because of their opposite signs,

it becomes

$$aaa - bba$$
$$- bca$$
$$- cca = - bbc$$
$$- bcc$$

the binomial equation which was sought.

Thus the required reduction of the proposed trinomial equation to the binomial which was sought is, in this way, accomplished.

PROBLEM 3

To reduce the trinomial equation

$$aaa - baa + bca$$
$$- caa - bda$$
$$+ daa - cda = -bcd$$

to the binomial

$$aaa - bbaa$$
$$- bcaa$$
$$- ccaa = - bccc$$
$$\overline{b + c} \qquad \overline{b + c}$$

by removing the first degree a.

Let us put $bc = bd + cd$

Then in the proposed equation, the first degree a is expunged on account of the opposite signs of the terms.

There will remain

$$aaa - baa$$
$$- caa$$
$$+ daa = -bcd$$

the part of the equation which still must be reduced.

Putting $bc = bd + cd$ it is true that $\frac{bc}{b+c} = d$.

And so, in the part of the equation which remains, let d be changed into $\frac{bc}{b+c}$.

From which we get

$$aaa - baa$$
$$- caa$$
$$+ \frac{bcaa}{b + c} \, {}^{3} = - \frac{bbcc}{b + c}$$

Let the other terms in baa and caa be expressed with the common divisor $b + c$.

From which it follows that

$$aaa - bbaa$$
$$- bcaa$$
$$- bcaa$$
$$- ccaa$$
$$+ \frac{bcaa}{b + c} = - \frac{bbcc}{b + c}$$

31 Removing the redundant terms of opposite signs,

it thus finally becomes

$$aaa - bbaa$$
$$- bcaa$$
$$- ccaa = - \frac{bbcc}{b + c}$$
$$\overline{b + c}$$

the binomial equation which was designated for the reduction.

Thus the required reduction of the proposed equation to that which was designated is, in this way, accomplished.

PROBLEM 4

To reduce the trinomial equation
$$\begin{aligned} aaa + baa + bca \\ + caa - bda \\ - daa - cda = +bcd \end{aligned}$$

to the binomial
$$\begin{aligned} aaa - bba \\ - bca \\ - cca = + bbc \\ + bcc, \end{aligned}$$
that is, by removing the second degree aa.

Let us put $b + c = d$.

And in the trinomial equation which is to be reduced, let d be changed into $b + c$.

From which it follows that
$$\begin{aligned} aaa + baa + bca \\ + caa - bba \\ - baa - bca \\ - caa - bca \\ - cca = + bbc \\ + bcc \end{aligned}$$

And so, removing the terms which are redundant on account of their opposite signs, it becomes
$$\begin{aligned} aaa - bba \\ - bca \\ - cca = + bbc \\ + bcc, \end{aligned}$$
the binomial equation which was sought.

Thus the required reduction of the proposed trinomial equation to the binomial which was sought is, in this way, accomplished.

PROBLEM 5

To reduce the trinomial
$$\begin{aligned} aaa + baa + bca \\ + caa - bda \\ - daa - cda = +bcd \end{aligned}$$

to the binomial
$$\begin{aligned} aaa + bbaa \\ + bcaa \\ + ccaa = + bbcc, \\ \overline{b + c} \qquad \overline{b + c} \end{aligned}$$
that is, by removing the first degree a.

Let us put $bc = bd + cd$.

Then in the equation which is to be reduced, the first degree a is removed on account of the opposite signs of the terms.

There will remain
$$\begin{aligned} + baa \\ + caa \\ - daa = +bcd \end{aligned}$$
the part of the equation which is still to be reduced.

From the substitution $bc = bd + cd$ we get $\frac{bc}{b+c} = d$.

And so in the remaining part of the equation and in the terms which involve d, let d be changed into $\frac{bc}{b+c}$

From which it follows that
$$\begin{aligned} aaa + baa \\ + caa \\ - bcaa = + bbcc \\ \overline{b + c} \qquad \overline{b + c} \end{aligned}$$

Let the other terms in baa and caa be expressed with the common divisor $b + c$.

Which gives

$$\begin{array}{r} aaa + bbaa \\ + bcaa \\ + bcaa \\ + ccaa \\ - bcaa \\ \hline b + c \end{array} = \frac{+ bbcc}{b + c}$$

Let the redundant terms of opposite signs be removed.

Hence it becomes

$$\begin{array}{r} aaa + bbaa \\ + bcaa \\ + ccaa \\ \hline b + c \end{array} = \frac{+ bbcc}{b + c},$$ the binomial equation which was

sought.

Thus the required reduction of the proposed trinomial equation to the binomial which was sought is, in this way, accomplished.

33

PROBLEM 6[4]

To reduce the trinomial equation $aaa - 3baa + 3bba = bbb - ccc$ to the binomial $aaa + 3bca = bbb - ccc$, that is, by removing the second degree aa.

Let us take $b - a = +c$, the equated root of generation.

Therefore $\overline{-a + b} \big|_{3ba} = \overline{+c} \big|_{3ba}$

But $\overline{-a + b} \big|_{3ba} = -3baa + 3bba$.

And $\overline{+c} \big|_{3ba} = +3bca$.

Therefore $-3baa + 3bba = +3bca$

Therefore $aaa - 3baa + 3bba = aaa + 3bca$.

But $aaa - 3baa + 3bba = bbb - ccc$.

And this is precisely the proposed trinomial equation.

Therefore $aaa + 3bca = bbb - ccc$, and this is the binomial equation which was sought.

Thus by putting $b - a = +c$ the required reduction of the proposed trinomial equation to the binomial which was sought is accomplished.

PROBLEM 7

To reduce the trinomial equation $aaa + 3baa + 3bba = -bbb + ccc$ to the binomial $aaa + 3bca = -bbb + ccc$, that is, by removing the second degree aa.

Let us take $a + b = +c$, the equated root of generation.[5]

Therefore $\overline{+a + b} \big|_{3ba} = \overline{+c} \big|_{3ba}$

But $\overline{+a + b} \big|_{3ba} = +3baa + 3bba$

And $\begin{array}{c}+c\\ \overline{3ba}\end{array}\bigg| = +3bca$

Therefore $+3baa + 3bba = +3bca$

Therefore $aaa + 3baa + 3bba = aaa + 3bca$

But $aaa + 3baa + 3bba = -bbb + ccc.$

And this is precisely the proposed trinomial equation.

Therefore $aaa + 3bca = -bbb + ccc$, and this is the binomial equation which **34** was sought.

Thus by putting $a + b = +c$ the required reduction of the proposed trinomial equation to the binomial equation which was sought is accomplished.

PROBLEM 8

To reduce the trinomial equation $aaa - 3baa + 3bba = +bbb + ccc$ to the binomial $aaa - 3bca = +bbb + ccc$, that is, by removing the second degree aa.

Let us take $a - b = +c$, the equated root of generation.[6]

Therefore $\begin{array}{c}-a + b\\ \overline{3ba}\end{array}\bigg| = \begin{array}{c}-c\\ \overline{3ba}\end{array}\bigg|$

But $\begin{array}{c}-a + b\\ \overline{3ba}\end{array}\bigg| = -3baa + 3bba$

And $\begin{array}{c}-c\\ \overline{3ba}\end{array}\bigg| = -3bca.$

Therefore $-3baa + 3bba = -3bca$

Therefore $aaa - 3baa + 3bba = aaa - 3bca$

But $aaa - 3baa + 3bba = +bbb + ccc$

And this is precisely the proposed trinomial equation.

Therefore $aaa - 3bca = +bbb + ccc$, and this is the binomial equation which was sought.

Thus by putting $a - b = +c$, the demanded reduction of the proposed trinomial equation binomial equation which was sought is accomplished.

Reductions of Canonical Equations of the Biquadratic Order

PROBLEM 9

To reduce the quadrinomial equation
$$\begin{aligned}
aaaa - baaa + bcaa\\
- caaa + bdaa\\
- daaa + cdaa - bcda\\
+ faaa - bfaa + bcfa\\
- cfaa + bdfa\\
- dfaa + cdfa = +bcdf
\end{aligned}$$

35 to the trinomial

$$
\begin{aligned}
aaaa &- bbaa + bbca \\
&- ccaa + bbda \\
&- ddaa + bcca \\
&- bcaa + ccda \\
&- bdaa + bdda \\
&- cdaa + cdda \\
&+ 2bcda = + bbcd \\
&\qquad\qquad\; + bccd \\
&\qquad\qquad\; + bcdd,
\end{aligned}
$$

that is, by removing

the third degree aaa.

Let us put $b + c + d = f$.

And in the quadrinomial equation which was proposed for reduction, let f be changed into $b + c + d$.

From which we get

$$
\begin{aligned}
aaaa &- baaa + bcaa \\
&- caaa + bdaa \\
&- daaa + cdaa - bcda \\
&+ baaa - bbaa + bcba \\
&+ caaa - bcaa + bcca \\
&+ daaa - bdaa + bcda \\
&- cbaa + bdba \\
&- ccaa + bdca \\
&- cdaa + bdda \\
&- dbaa + cdba \\
&- dcaa + cdca \\
&- ddaa + cdda = + bcdb \\
&\qquad\qquad\qquad + bcdc \\
&\qquad\qquad\qquad + bcdd
\end{aligned}
$$

Remove the terms which are redundant because of their opposite signs.

Hence it becomes

$$
\begin{aligned}
aaaa &- bbaa + bbca \\
&- ccaa + bbda \\
&- ddaa + bcca \\
&- bcaa + ccda \\
&- bdaa + bdda \\
&- cdaa + cdda \\
&+ 2bcda = + bbcd \\
&\qquad\qquad\; + bccd \\
&\qquad\qquad\; + bcdd
\end{aligned}
$$

But this is the required trinomial equation. Thus the required reduction of the proposed equation to that which was sought is, in this way, accomplished.

36   PROBLEM 10

To reduce the quadrinomial equation

$$
\begin{aligned}
aaaa &- baaa + bcaa \\
&- caaa + bdaa \\
&- daaa + cdaa - bcda \\
&+ faaa - bfaa + bcfa \\
&\qquad\quad - cfaa + bdfa \\
&\qquad\quad - dfaa + cdfa = + bcdf
\end{aligned}
$$

$$
\begin{array}{l}
aaaa - bbaaa + bbcca \\
\quad\ \ - ccaaa + bbdda \\
\quad\ \ - ddaaa + ccdda \\
\quad\ \ - bcaaa + bcdda \\
\end{array}
$$

to the trinomial $\dfrac{\begin{array}{l} - bdaaa + bccda \\ - cdaaa + bbcda \\ \hline b + c + d\ b + c + d \end{array}}{} = \dfrac{\begin{array}{l}+ bbccd \\ + bbcdd \\ + bccdd, \\ \hline b + c + d\end{array}}{}$ that is, by removing the

second degree aa.

Let us put $bc + bd + cd = bf + cf + df$.

Then in the proposed equation, the second degree aa is removed through the opposite signs of the terms.

There will remain $\begin{array}{l} aaaa - baaa - bcda \\ \quad\ \ - caaa + bcfa \\ \quad\ \ - daaa + bdfa \\ \quad\ \ + faaa + cdfa = +bcdf \end{array}$ the part of the equation

which still must be reduced.

From the substitution $bc + bd + cd = bf + cf + df$ we get $\dfrac{bc+bd+cd}{b+c+d} = f$.

And so in the remaining part of the equation and in the terms which involve f, let f be changed into $\dfrac{bc+bd+cd}{b+c+d}$

From which it follows that $\dfrac{\begin{array}{l} aaaa - baaa\quad - bcda \\ \quad\ \ - caaa\quad + bbcca \\ \quad\ \ - daaa\quad + bbcda \\ \quad\ \ + bcaaa\ + bccda \\ \quad\ \ + bdaaa\ + bbcda \\ \quad\ \ + cdaaa\ + bbdda \\ \hline b + c + d\ + bcdda \\ \quad\quad\ \ + bccda \\ \quad\quad\ \ + bcdda \\ \quad\quad\ \ + ccdda \end{array}}{} = \dfrac{\begin{array}{l} + bbccd \\ + bbcdd \\ + bccdd \\ \hline b + c + d \end{array}}{}$

Let the remaining terms in $baaa$, $caaa$, $daaa$ and $bcda$ be expressed with the **37** common divisor $b + c + d$.

Which gives $\dfrac{\begin{array}{l} aaaa - bbaaa\ - bbcda \\ \quad\ \ - bcaaa\ - bccda \\ \quad\ \ - bdaaa\ - bcdda \\ \quad\ \ - bcaaa\ + bbcca \\ \quad\ \ - ccaaa\ + bbcda \\ \quad\ \ - dcaaa\ + bccda \\ \quad\ \ - bdaaa\ + bbcda \\ \quad\ \ - cdaaa\ + bbdda \\ \quad\ \ - ddaaa\ + bcdda \\ \quad\ \ + bcaaa\ + bccda \\ \quad\ \ + bdaaa\ + bcdda \\ \quad\ \ + cdaaa\ + ccdda \\ \hline b + c + d\ b + c + d \end{array}}{} = \dfrac{\begin{array}{l} + bbccd \\ + bbcdd \\ + bccdd \\ \hline b + c + d \end{array}}{}$

Remove the terms which are redundant because of their opposite signs.

Hence it becomes

$$\begin{array}{l} aaaa - bbaaa + bbcca \\ \quad - bcaaa + bbcda \\ \quad - ccaaa + bccda \\ \quad - bdaaa + bbdda \\ \quad - cdaaa + bcdda \\ \underline{\quad - ddaaa + ccdda} = + bbccd \\ \overline{b+c+d}\; \overline{b+c+d} \;\; + bbcdd \\ \qquad\qquad\qquad\qquad + bccdd \\ \qquad\qquad\qquad\qquad \overline{b+c+d} \end{array}$$

But this is the required trinomial equation.

Thus the required reduction of the proposed quadrinomial equation to the trinomial equation which was sought is, in this way, accomplished.

PROBLEM 11

To reduce the quadrinomial equation

$$\begin{array}{l} aaaa - baaa + bcaa \\ \quad - caaa + bdaa \\ \quad - daaa + cdaa - bcda \\ \quad + faaa - bfaa + bcfa \\ \qquad\qquad - cfaa + bdfa \\ \qquad\qquad - dfaa + cdfa = +bcdf \end{array}$$

38 to the trinomial

$$\begin{array}{l} aaaa - bbcaaa \\ \quad - bbdaaa \quad + bbccaa \\ \quad - bccaaa \quad + bbddaa \\ \quad - bddaaa \quad + ccddaa \\ \quad - ccdaaa \quad + bcddaa \\ \quad - cddaaa \quad + bccdaa \\ \underline{\quad - 2bcdaaa \quad + bbcdaa} = \underline{bbccdd} \\ \overline{bc+bd+cd}\; \overline{bc+bd+cd} \quad \overline{bc+bd+cd} \end{array} ,$$

that is, by removing the first degree a.

Let us put $bcd = bcf + bdf + cdf$

Then in the proposed quadrinomial equation, the first degree a is removed through the opposite signs of the terms.

There will remain

$$\begin{array}{l} aaaa - baaa + bcaa \\ \quad - caaa + bdaa \\ \quad - daaa + cdaa \\ \quad + faaa - bfaa \\ \qquad\qquad - cfaa \\ \qquad\qquad - dfaa = +bcdf, \end{array}$$

the part of the equation

which still must be reduced.

From the substitution $bcd = bcf + bdf + cdf$ we get $\frac{bcd}{bc+bd+cd} = f$.

And so in the remaining part of the equation and in the terms which involve f, first let f be changed into $\frac{bcd}{bc+bd+cd}$.

Secondly, express the remaining terms with the common divisor $bc + bd + cd$.

Thirdly, remove the terms which are redundant because of their opposite signs.

Having done this (as in Probl. 10) we get

$$
\begin{array}{lll}
aaaa - bbcaaa & & \\
\quad - bbdaaa & + bbccaa & \\
\quad - bccaaa & + bbddaa & \\
\quad - bddaaa & + ccddaa & \\
\quad - ccdaaa & + bcddaa & \\
\quad - cddaaa & + bccdaa & \\
\quad - 2bcdaaa & + bbcdaa & = + bbccdd \\
\hline
bc + bd + cd & bc + bd + cd & bc + bd + cd
\end{array}
$$

But this is the required trinomial equation, in which the first degree a has been removed.

And thus the required reduction is accomplished.

PROBLEM 12   **39**

To reduce the quadrinomial equation

$$
\begin{array}{l}
aaaa + baaa + bcaa \\
\quad + caaa + bdaa \\
\quad + daaa + cdaa + bcda \\
\quad - faaa - bfaa - bcfa \\
\quad\qquad - cfaa - bdfa \\
\quad\qquad - dfaa - cdfa = +bcdf
\end{array}
$$

to the trinomial

$$
\begin{array}{l}
aaaa - bbaa - bbca \\
\quad - ccaa - bbda \\
\quad - ddaa - bcca \\
\quad - bcaa - ccda \\
\quad - bdaa - bdda \\
\quad - cdaa - cdda \\
\quad\qquad - 2bcda = + bbcd \\
\quad\qquad\qquad\qquad + bccd \\
\quad\qquad\qquad\qquad + bcdd,
\end{array}
$$

that is, by removing the third degree aaa.

Let us put $b + c + d = f$.

And in the quadrinomial equation which is proposed for reduction, let f be changed into $b + c + d$.

From which it follows that

$$
\begin{array}{l}
aaaa + baaa + bcaa \\
\quad + caaa + bdaa \\
\quad + daaa + cdaa + bcda \\
\quad - baaa - bbaa - bcba \\
\quad - caaa - bcaa - bcca \\
\quad - daaa - bdaa - bcda \\
\quad\qquad - cbaa - bdba \\
\quad\qquad - ccaa - bdca \\
\quad\qquad - cdaa - bdda \\
\quad\qquad - dbaa - cdba \\
\quad\qquad - dcaa - cdca \\
\quad\qquad - ddaa - cdda = + bcdb \\
\quad\qquad\qquad\qquad\qquad + bcdc \\
\quad\qquad\qquad\qquad\qquad + bcdd
\end{array}
$$

Remove the terms which are redundant because of their opposite signs.

Hence it becomes
$$\begin{aligned}
aaaa &- bbaa - bbca \\
&- bcaa - bcca \\
&- ccaa - bbda \\
&- dbaa - bdda \\
&- dcaa - ccda \\
&- ddaa - cdda \\
&\quad\ - 2bcda = + bbcd \\
&\qquad\qquad\quad + bcdc \\
&\qquad\qquad\quad + bcdd,
\end{aligned}$$

the trinomial equation which was sought.

Thus the required reduction of the proposed equation to that which was sought is, in this way, accomplished.

40

PROBLEM 13

To reduce the quadrinomial equation
$$\begin{aligned}
aaaa &+ baaa + bcaa \\
&+ caaa + bdaa \\
&+ daaa + cdaa + bcda \\
&- faaa - bfaa - bcfa \\
&\qquad\quad - cfaa - bdfa \\
&\qquad\quad - dfaa - cdfa = +bcdf
\end{aligned}$$

to the trinomial
$$\frac{\begin{aligned}
aaaa &+ bbaaa - bbcca \\
&+ ccaaa - bbdda \\
&+ ddaaa - ccdda \\
&+ bcaaa - bcdda \\
&+ bdaaa - bccda \\
&+ cdaaa - bbcda = + bbccd
\end{aligned}}{b+c+d\,\,b+c+d} \quad \frac{\begin{aligned} \\ + bbcdd \\ + bccdd \end{aligned}}{b+c+d,}$$

that is, by removing the second degree aa.

Let us put $bc + bd + cd = bf + cf + df$.

Then in the proposed quadrinomial equation, the second degree aa is removed through the opposite signs of the terms.

There will remain
$$\begin{aligned}
aaaa &+ baaa + bcda \\
&+ caaa - bcfa \\
&+ daaa - bdfa \\
&- faaa - cdfa = +bcdf,
\end{aligned}$$

the part of the equation which still must be reduced.

From the substitution $bc + bd + cd = bf + cf + df$ we get $\frac{bc+bd+cd}{b+c+d} = f$.

And so in the remaining part of the equation and in the terms which involve f, first let f be changed into $\frac{bc+bd+cd}{b+c+d}$

Secondly, express the remaining terms with a common divisor.

Thirdly, remove the terms which are redundant because of their opposite signs.

Having done this (as in Probl. 10) we get

$$\frac{aaaa + bbaaa - bbcca}{b+c+d} \frac{+ ccaaa - bbdda}{b+c+d} \begin{array}{l} + ddaaa - ccdda \\ + bcaaa - bcdda \\ + bdaaa - bccda \\ + cdaaa - bbcda = + bbccd \\ \hline + bbcdd \\ + bccdd \\ \hline b+c+d \end{array}$$

But this is the required trinomial equation.   **41**

And thus the required reduction is accomplished.

PROBLEM 14

To reduce the quadrinomial equation

$$\begin{array}{l} aaaa + baaa + bcaa \\ + caaa + bdaa \\ + daaa + cdaa + bcda \\ - faaa - bfaa - bcfa \\ - cfaa - bdfa \\ - dfaa - cafa = + bcdf \end{array}$$

to the trinomial

$$\frac{aaaa + bbcaaa}{bc+bd+cd} \begin{array}{l} + bbdaaa + bbccaa \\ + bccaaa + bbddaa \\ + ccdaaa + ccddaa \\ + bddaaa + bcddaa \\ + cddaaa + bccdaa \\ + 2bcdaaa + bbcdaa = + bbccdd, \end{array}$$

that is, by

removing the first degree a.

Let us put $bcd = bcf + bdf + cdf$.

Then in the proposed quadrinomial equation, the first degree a is removed through the opposite signs of the terms.

There will remain

$$\begin{array}{l} aaaa + baaa + bcaa \\ + caaa + bdaa \\ + daaa + cdaa \\ - faaa - bbaa \\ - cfaa \\ - dfaa = + bcdf. \end{array}$$

the part of the proposed

equation which still must be reduced.

From the substitution $bcd = bcf + bdf + cdf$ we get $\frac{bcd}{bc+bd+cd} = f$.

And so in the remaining part of the equation and in the terms which involve f, first let f be changed into $\frac{bcd}{bc+bd+cd}$

Secondly, express the remaining terms with a common divisor.

Thirdly, remove the terms which are redundant because of their opposite signs.

42 Having done this (as in Probl. 10), we get

$$aaaa + bbcaaa$$
$$+ bbdaaa \quad + bbccaa$$
$$+ bccaaa \quad + bbddaa$$
$$+ ccdaaa \quad + ccddaa$$
$$+ bddaaa \quad + bcddaa$$
$$+ cddaaa \quad + bccdaa$$
$$\underline{+ 2bcdaaa \quad + bbcdaa} \qquad = \; + bbccdd,$$
$$\overline{bc + bd + cd} \; \overline{bc + bd + cd} \qquad \overline{bc + bd + cd}$$

the required trinomial equation in which the first degree a is removed.

And in this way the required reduction is accomplished.

PROBLEM 15

To reduce the quadrinomial equation

$$aaaa - baaa + bcaa$$
$$- caaa - bdaa$$
$$+ daaa - cdaa + bcda$$
$$+ faaa - bfaa + bcfa$$
$$- cfaa - bdfa$$
$$+ dfaa - cdfa = -bcdf$$

to the trinomial

$$aaaa + bdaa + bbca$$
$$+ cdaa + bcca$$
$$- bbaa + bdda$$
$$- bcaa + cdda$$
$$- ccaa - bbda$$
$$- ddaa - ccda$$
$$- 2bcda = - bbcd$$
$$- bccd$$
$$+ bcdd,$$

that is, removing the third

degree aaa.

If it were the case that $b + c = d + f$, then it will be true that $b + c - d = f$.

And so, let us put $b + c - d = f$.

And in the quadrinomial equation proposed for reduction, first let f be changed into $b + c - d$.

Secondly, remove the terms which are redundant because of their opposite signs.

Having done this (as in Probl. 10), we get

$$aaaa + bdaa + bbca$$
$$+ cdaa + bcca$$
$$- bbaa + bdda$$
$$- bcaa + cdda$$
$$- ccaa - bbda$$
$$- ddaa - ccda$$
$$- 2bcda = - bbcd$$
$$- bccd$$
$$+ bcdd,$$

the required trinomial equation, in which the third degree aaa is removed.

And in this way the required reduction is accomplished.

43

PROBLEM 16

To reduce the quadrinomial equation

$$aaaa - baaa + bcaa$$
$$- caaa - bdaa$$
$$+ daaa - cdaa + bcda$$
$$+ faaa - bfaa + bcfa$$
$$- cfaa - bdfa$$
$$+ dfaa - cdfa = -bcdf$$

to the trinomial

$$\frac{\begin{aligned}aaaa &- bbaaa + bbcca\\ &- bcaaa + bbdda\\ &- ccaaa + bcdda\\ &- ddaaa + ccdda\\ &+ bdaaa - bbcda\\ &+ cdaaa - bccda\end{aligned}}{\overline{b+c-d}\,\overline{b+c-d}} = \frac{\begin{aligned}- bbccd\\ + bbcdd\\ + bccdd\end{aligned}}{b+c-d},$$

that is, by removing the second degree aa.

If it were the case that $bc + df = bd + cd + bf + cf$, i.e., $bc - bd - cd = bf + cf - df$ then it will be true that $\frac{bc-bd-cd}{b+c-d} = f$.

And so, let us put $\frac{bc-bd-cd}{b+c-d} = f$.

And in the quadrinomial equation which is proposed for reduction, first let f be changed into $\frac{bc-bd-cd}{b+c-d}$

Secondly, express the remaining terms with the common divisor $b + c - d$.

Thirdly, remove the terms which are redundant because of their opposite signs.

Having done this (as in Probl. 9), we get

$$\frac{\begin{aligned}aaaa &- bbaaa + bbcca\\ &- bcaaa + bbdda\\ &- ccaaa + bcdda\\ &- ddaaa + ccdda\\ &+ bdaaa - bbcda\\ &+ cdaaa - bccda\end{aligned}}{\overline{b+c-d}\,\overline{b+c-d}} = \frac{\begin{aligned}- bbccd\\ + bbcdd\\ + bccdd\end{aligned}}{b+c-d},$$

the required trinomial equation, in which the second degree aa is removed.

And in this way the required reduction is accomplished.

PROBLEM 17[7]

The quadrinomial equation proposed above is also reducible to this trinomial:

$$\frac{\begin{aligned}aaaa &+ bbaaa - bbcca\\ &+ bcaaa - bcdda\\ &+ ccaaa - bbdda\\ &+ ddaaa - ccdda\\ &- bdaaa + bbcda\\ &- cdaaa + bccda\end{aligned}}{\overline{d-b-c}\,\overline{d-b-c}} = \frac{\begin{aligned}- bbcdd\\ - bccdd\\ + bbccd.\end{aligned}}{d-b-c}$$

that is, by the removal of the second degree aa putting $f = \frac{bd+cd-bc}{d-b-c}$, and (as above) changing f into $\frac{bd+cd-bc}{d-b-c}$, then expressing in terms of the the common divisor $d - b - c$, and removing the terms which are redundant on account of their opposite signs, after the model of reduction in Problem 9.

Problem 18

$$aaaa - baaa + bcaa$$
$$- caaa - bdaa$$
To reduce the quadrinomial equation
$$+ daaa - cdaa + bcda$$
$$+ faaa - bfaa + bcfa$$
$$- cfaa - bdfa$$
$$+ dfaa - cdfa = -bcdf$$

to the trinomial

$$\begin{array}{l}
aaaa + bbcaaa \\
\quad + bccaaa \quad - bbccaa \\
\quad + bddaaa \quad - bbddaa \\
\quad + cddaaa \quad - bcddaa \\
\quad - bbdaaa \quad - ccddaa \\
\quad - ccdaaa \quad + bbcdaa \\
\quad - 2bcdaaa \quad + bccdaa \quad = -bbccdd \\
\hline
\overline{bd + cd - bc}\ \overline{bd + cd - bc} \quad \overline{bd + cd - bc,}
\end{array}$$

that is, by

removal of the first degree a.

If it were the case that $bcd + bcf = bdf + cdf$, i.e., $bcd = bdf + cdf - bcf$.
then it will be true that $\frac{bcd}{bd+cd-bc} = f$.

45 And so, let us put $\frac{bcd}{bd+cd-bc} = f$.

And in the quadrinomial equation which is proposed for reduction, and in the terms which involve f, first let f be changed into $\frac{bcd}{bd+cd-bc}$

Secondly, express the remaining terms with the common divisor $bd + cd - bc$.

Thirdly, remove the terms which are redundant on account of their opposite signs.

Having done this (as in Probl. 9), we get

$$\begin{array}{l}
aaaa + bbcaaa \\
\quad + bccaaa \quad - bbccaa \\
\quad + bddaaa \quad - bbddaa \\
\quad + cddaaa \quad - bcddaa \\
\quad - bbdaaa \quad - ccddaa \\
\quad - ccdaaa \quad + bbcdaa \\
\quad - 2bcdaaa \quad + bccdaa \quad = -bbccdd \\
\hline
\overline{bd + cd - bc}\ \overline{bd + cd - bc} \quad \overline{bd + cd - bc}
\end{array}$$

the required trinomial equation in which the first degree a is removed.

And in this way the required reduction is accomplished.

Problem 19[8]

$$aaaa - baaa + bcaa$$
$$- caaa - bdaa$$
To reduce the quadrinomial equation
$$+ daaa - cdaa + bcda$$
$$+ faaa - bfaa + bcfa$$
$$- cfaa - bdfa$$
$$+ dfaa - cdfa = -bcdf$$

by putting $b + c = d + f$ to the binomial

$$\begin{array}{l}
aaaa - bbba \\
\quad - bbca \\
\quad - bcca \\
\quad - ccca = -bbbc \\
\quad\quad\quad\quad - bbcc \\
\quad\quad\quad\quad - bccc,
\end{array}$$

that is, by removing degrees aa and aaa.

PROBLEM 20

To reduce the quadrinomial equation

$$aaaa - baaa + bcaa$$
$$- caaa - bdaa$$
$$+ daaa - cdaa + bcda$$
$$+ faaa - bfaa + bcfa$$
$$- cfaa - bdfa$$
$$+ dfaa - cdfa = -bcdf$$

by putting $bc + df = bd + cd + bf + cf$

to the binomial
$$\frac{aaaa - bbbaa - bbcaaa - bccaaa - cccaaa}{bb + bc + cc} = \frac{-bbbccc}{bb + bc + cc},$$
that is, by removing the **46** degrees a and aaa.

PROBLEM 21

To reduce the quadrinomial equation

$$aaaa - baaa + bcaa$$
$$- caaa - bdaa$$
$$+ daaa - cdaa + bcda$$
$$+ faaa - bfaa + bcfa$$
$$- cfaa - bdfa$$
$$+ dfaa - cdfa = -bcdf$$

by putting $d + f = b + c$

to the binomial $\dfrac{aaaa - bbaa - ccaa = -bbcc,}{}$ that is, by removing the degrees a and aaa.

Note

The three preceding binomial reductions in Problems 19, 20 and 21 correspond to the three trinomials above in problems 16, 17 and 18 (which are reduced from the same quadrinomial as is proposed here) in having the same explicatory roots b and c, as is demonstrated in propositions 32, 33 and 34, and 35, 36 and 37 of the Fourth Section. Since, however, their reductions have been recorded most obscurely in the manuscripts, they must await a more thorough enquiry.

General Corollary

In the reductions performed by the Problems of this Third Section, it is clear that there is no change made in the sought root a, nor in the other degrees, nor in the given terms b, c and d.

Summary of the Reduced Canonical Equations whose Reductions have been presented in this Third Section

Quadratic

1. $aa = bb$

Cubics

1. $aaa - bba$
   $- bca$
   $- cca = - bbc$
   $\qquad\qquad - bcc$

47 2. $aaa - bbaa$
   $- bcaa$
   $\dfrac{- ccaa}{b + c} = \dfrac{- bbcc}{b + c}$

3. $aaa - bba$
   $- bca$
   $- cca = + bbc$
   $\qquad\qquad + bcc$

4. $aaa + bbaa$
   $+ bcaa$
   $\dfrac{+ ccaa}{b + c} = \dfrac{+ bbcc}{b + c}$

5. $aaa + 3bca = +bbb - ccc$

6. $aaa + 3bca = -bbb + ccc$

7. $aaa - 3bca = +bbb + ccc$

Biquadratics

1. $aaaa - bbaa + bbca$
   $\quad - ccaa + bbda$
   $\quad - ddaa + bcca$
   $\quad - bcaa + ccda$
   $\quad - bdaa + bdda$
   $\quad - cdaa + cdda$
   $\qquad\qquad + 2bcda = + bbcd$
   $\qquad\qquad\qquad\qquad + bccd$
   $\qquad\qquad\qquad\qquad + bcdd$

2. $aaaa - bbaaa \quad + bbcca$
   $\quad - ccaaa \quad + bbdda$
   $\quad - ddaaa \quad + ccdda$
   $\quad - bcaaa \quad + bcdda$
   $\quad - bdaaa \quad + bccda$
   $\dfrac{- cdaaa}{b + c + d} \quad \dfrac{+ bbcda}{b + c + d} \quad = \quad + bbccd$
   $\qquad\qquad\qquad\qquad\qquad\qquad + bbcdd$
   $\qquad\qquad\qquad\qquad\qquad\qquad \dfrac{+ bccdd}{b + c + d}$

3.
$$aaaa - \frac{bbcaaa + bbdaaa + bccaaa + bddaaa + ccdaaa + cddaaa + 2bcdaaa}{bc+bd+cd} + \frac{bbccaa + bbddaa + ccddaa + bcddaa + bccdaa + bbcdaa}{bc+bd+cd} = \frac{+bbccdd}{bc+bd+cd}$$

48

4.
$$aaaa - bbaa - ccaa - ddaa - bcaa - bdaa - cdaa - bbca - bbda - bcca - ccda - bdda - cdda - 2dcba = +bbcd + bccd + bcdd$$

5.
$$aaaa + \frac{bbaaa + ccaaa + ddaaa + bcaaa + bdaaa + cdaaa}{b+c+d} - \frac{bbcca + bbdda + ccdda + bcdda + bccda + bbcda}{b+c+d} = \frac{+bbccd + bbcdd + bccdd}{b+c+d}$$

6.
$$aaaa + \frac{bbcaaa + bbdaaa + bccaaa + ccdaaa + bddaaa + cddaaa + 2bcdaaa}{bc+bd+cd} + \frac{bbccaa + bbddaa + ccddaa + bcddaa + bccdaa + bbcdaa}{bc+bd+cd} = \frac{+bbccdd}{bc+bd+cd}$$

7.
$$aaaa - bdaa - cdaa + bbaa + bcaa + ccaa + ddaa - bbca - bcca - bdda - cdda + bbda + ccda + 2bcda = +bbcd + bccd - bcdd$$

9

8.
$$
\begin{array}{c}
aaaa - bbaaa + bbcca \\
- bcaaa + bbdda \\
- ccaaa + bcdda \\
- ddaaa + ccdda \\
+ bdaaa - bbcda \\
+ cdaaa - bccda = - bbccd \\
\overline{b + c - d} \; \overline{b + c - d} \quad + bbcdd \\
+ bccdd \\
\overline{b + c + d}
\end{array}
$$

49 9.
$$
\begin{array}{c}
aaaa + bbaaa - bbcca \\
+ bcaaa - bbdda \\
+ ccaaa - bcdda \\
+ ddaaa - ccdda \\
- bdaaa + bbcda \\
- cdaaa + bccda = - bbcdd \\
\overline{d - b - c} \; \overline{d - b - c} \quad - bccdd \\
+ bbccd \\
\overline{d - b - c}
\end{array}
$$

10.
$$
\begin{array}{c}
aaaa + bbcaaa \\
+ bccaaa \qquad - bbcca \\
+ bddaaa \qquad - bbdda \\
+ cddaaa \qquad - bcdda \\
- bbdaaa \qquad - ccdda \\
- ccdaaa \qquad + bbcda \\
- 2bcdaaa \qquad + bccda \qquad = - bbccdd \\
\overline{bd + cd - bc} \; \overline{bd + cd - bc} \quad \overline{bd + cd - bc}
\end{array}
$$

11.
$$
\begin{array}{c}
aaaa - bbba \\
- bbca \\
- bcca \\
- ccca = - bbbc \\
- bbcc \\
- bccc
\end{array}
$$

12.
$$
\begin{array}{c}
aaaa - bbbaaa \\
- bbcaaa \\
- bccaaa \\
- cccaaa \qquad = - bbbccc \\
\overline{bb + bc + cc} \quad \overline{bb + bc + cc}
\end{array}
$$

13.
$$
\begin{array}{c}
aaaa - bbaa \\
- ccaa = -bbcc
\end{array}
$$

Summary of Certain Canonical Equations, Set Out in Such a Way that the Generation of Others of Higher Degree may Easily be Perceived[10]

$$+bc = +ba$$
$$+ca - aa$$

$$+ bbc = +bba$$
$$+ bcc \quad + bca$$
$$+ cca - aaa$$

$$+ bbbc = + bbba$$
$$+ bbcc \quad + bbca$$
$$+ bccc \quad + bcca$$
$$+ ccca - aaaa$$

$$+ bbbbc = + bbbba$$
$$+ bbbcc \quad + bbbca$$
$$+ bbccc \quad + bbcca$$
$$+ bcccc \quad + bccca$$
$$+ cccca - aaaaa$$

50

And so to infinity by the same method.

$$+ bbcc = + bbaa$$
$$\overline{b + c} \quad + bcaa$$
$$+ ccaa - aaa$$
$$\overline{b + c}$$

$$+ bbbcc = + bbbaa$$
$$+ bbccc \quad + bbcaa$$
$$\overline{b + c} \quad + bccaa$$
$$+ cccaa - aaaa$$
$$\overline{b + c}$$

$$+ bbbbcc = + bbbbaa$$
$$+ bbbccc \quad + bbbcaa$$
$$+ bbcccc \quad + bbccaa$$
$$\overline{b + c} \quad + bcccaa$$
$$+ ccccaa - aaaaa$$
$$\overline{b + c}$$

And so to infinity by the same method.

$$\frac{+ bbbccc}{bb + bc + cc} = + bbbaaa$$
$$+ bbcaaa$$
$$+ bccaaa$$
$$\frac{+ cccaaa}{bb + bc + cc} - aaaaa$$

$$\begin{array}{ll} + bbbbccc & = + bbbbaaa \\ + bbbcccc & + bbbcaaa \\ \overline{bb + bc + cc} & + bbccaaa \\ & + bcccaaa \\ & \underline{+ ccccaaa - aaaaa} \\ & \overline{bb + bc + cc} \end{array}$$

And so to infinity in other cases by the same method.

$$\begin{array}{ll} \dfrac{+bbbbccc}{bbb + bbc + bcc + ccc} & = + bbbbaaaa \\ & + bbbcaaaa \\ & + bbccaaaa \\ & + bcccaaaa \\ & \underline{+ ccccaaaa} \qquad -aaaaa \\ & \overline{bbb + bbc + bcc + ccc} \end{array}$$

And so to infinity in other cases by the same method.

51 Another collection and series of Canonicals

$$\begin{array}{l} +bcd = + bca - baa \\ \qquad\quad + bda - caa \\ \qquad\quad + cda - daa + aaa \end{array}$$

$$\begin{array}{ll} + bbcd = + bbca \\ + cbcd & + bbda - bbaa \\ + dbcd & + ccba - ccaa \\ & + ccda - ddaa \\ & + ddba - bcaa \\ & + ddca - bdaa \\ & + 2bcda - cdaa + aaaa \end{array}$$

$$\begin{array}{ll} + bcbcd & = + bbcca \; - bbaaa \\ + bdbcd & + bbdda \; - ccaaa \\ + bdbcd & + ccdda \; - ddaaa \\ \overline{b + c + d} & + bbcda \; - bcaaa \\ & + cbcda \; - bdaaa \\ & \underline{+ dbcda \; - cdaaa + aaaa} \\ & \overline{b + c + d}\,\overline{b + c + d} \end{array}$$

$$\begin{array}{ll} +bcdbcd = + bbccaa & - bbcaaa \\ \qquad\quad + bbddaa & - bbdaaa \\ \qquad\quad + ccddaa & - ccbaaa \\ \qquad\quad + bbcdaa & - ccdaaa \\ \qquad\quad + cbcdaa & - ddbaaa \\ \qquad\quad \underline{+ dbcdaa \quad - ddcaaa} \\ \qquad\quad \overline{bc + bd + cd} - 2bcdaaa + aaaa \\ \qquad\qquad\quad \overline{bc + bd + cd} \end{array}$$

$$
\begin{array}{lll}
+\,bbcbcd & =+\,bbbcca & \\
+\,bbdbcd & +\,bbbdda & \\
+\,ccbbcd & +\,cccbba & -\,bbbaaa \\
+\,ccdbcd & +\,bbbcda & -\,cccaaa \\
+\,ddbbcd & +\,cccdda & -\,dddaaa \\
+\,ddcbcd & +\,cccbda & -\,bbcaaa \\
+\,2bcdbcd & +\,dddbba & -\,bbdaaa \\
\overline{b+c+d} & +\,dddcca & -\,ccbaaa \\
 & +\,dddbca & -\,ccdaaa \\
 & +\,2bcbcba & -\,ddbaaa \\
 & +\,2bdbcba & -\,ddcaaa \\
 & +\,2bdbcda & -\,bcdaaa & +\,aaaaa \\
 & \overline{b+c+d} & \overline{b+c+d}
\end{array}
$$

Section Four[1]

Determination of the Roots of Primary and Secondary Canonical Equations

PROPOSITION 1

b [2] is the [3] root of the equation $\begin{array}{c} aa - ba \\ + ca \end{array} = +bc$ equal to the sought root a.

For if b is put equal to the root a of the equation $\begin{array}{c} aa - ba \\ + ca \end{array} = +bb,$ then changing a into b we get $\begin{array}{c} bb - bb \\ + cb \end{array} = +cb.$

But the equality here is obvious.

Therefore, having put b equal to the root a[4], it is indeed [shown to be] equal, as was stated in the proposition.

And in the following Lemma it is demonstrated that there is no other root besides b equal to the root a of the equation[5].

LEMMA

Suppose another root of the equation equal to the root a and not equal to b could be given; let it be c (or anything else).[6]

Then putting $c = a$ it will be true that $\begin{array}{c} cc - bc \\ + cc \end{array} = +bc.$

Therefore $cc + cc = +bc + bc.$

Therefore $\dfrac{c+c}{c} = \dfrac{c+c}{b}$

Therefore $c = b$ which is contrary to the hypothesis.

Therefore it is not true that $c = a$ as was assumed. This can be demonstrated in a similar way for any other root apart from b.

Proposition 2

b or c are roots of the equation $\begin{aligned}aa-ba\\-ca\end{aligned}=-bc,$ equal to the sought root a.

For if b is put equal to the root a of the equation $\begin{aligned}aa-ba\\-ca\end{aligned}=-bc,$ then changing a into b we get $\begin{aligned}bb-bb\\-cb\end{aligned}=-bc$

53 But the equality here is obvious.

Therefore, having put b equal to the root a, it is indeed [shown to be] equal.

Likewise if c is put equal to the root a, then changing a into c we get $\begin{aligned}cc-bc\\-cc\end{aligned}=-bc.$

And the equality is also obvious here.

Therefore having put c equal to the root a, it is indeed [shown to be] equal.

Thus b and c are equal to the sought root a, as was stated in the proposition.
And in the following Lemma it is demonstrated that there is no other root besides b and c equal to the root a of the equation.

Lemma

Suppose another root of the equation equal to the root a and not equal to the roots b and c could be given; let it be d (or anything else).[7]

Then putting $d = a$ it will be true that $\begin{aligned}dd-bd\\-cd\end{aligned}=-bc.$

Therefore $dd - cd = +bd - bc$.

Therefore $\begin{array}{|c}+d-c\\\hline d\end{array} = \begin{array}{|c}+d-c\\\hline b\end{array}$

Therefore $d = b$ which is contrary to the hypothesis.

Or it will be true that $+dd - bd = +cd - bc$.

Therefore $\begin{array}{|c}+d-b\\\hline d\end{array} = \begin{array}{|c}+d-b\\\hline c\end{array}$

Therefore $d = c$ which is contrary to the hypothesis.

Therefore it is not true that $d = a$ as was assumed. This can be demonstrated in a similar way for any other root apart from b and c.

Proposition 3

d is the root of the equation $\begin{aligned}aaa+baa+bca\\+caa-bda\\-daa-cda\end{aligned}=+bcd,$ equal to the sought root a.

For if d is put equal to the root a of the equation $\begin{aligned}aaa+baa+bca\\+caa-bda\\-daa-cda\end{aligned}=+bcd,$

then changing a into d we get $\begin{aligned}ddd+bdd+bcd\\+cdd-bdd\\-ddd-cdd\end{aligned}=+bcd.$

54 But, once terms of opposite signs have been removed, the equality here is obvious.

Therefore having put d equal to the root a, it is indeed [shown to be] equal.

And in the following Lemma it is demonstrated that there is no other root besides d equal to the root a of the equation.

LEMMA

Suppose another root of the equation equal to the root a and not equal to the root d could be given; let it be b or c (or anything else).

Then putting $c = a$ it will be true that
$$\begin{aligned} ccc + bcc + bcc \\ + ccc - bdc \\ - dcc - cdc = +bcd \end{aligned}$$

And on rearranging the terms $+2ccc + 2bcc = +2ccd + 2bcd$.

Therefore
$$\frac{+cc + bc}{c} = \frac{+cc + bc}{d}$$

Therefore $c = d$ which is contrary to the hypothesis.

Therefore it is not true that $c = a$ as was assumed. And by precisely the same reasoning the same thing can be concluded about b or any other root apart from d.

PROPOSITION 4

c and d are the explicatory roots of the equation
$$\begin{aligned} aaa + baa - bca \\ - caa - bda \\ - daa + cda = -bcd \end{aligned}$$
equal to the sought root a.

For if c is put equal to the root a of the equation
$$\begin{aligned} aaa + baa - bca \\ - caa - bda \\ - daa + cda = -bcd \end{aligned}$$
then, changing a into c we get
$$\begin{aligned} ccc + bcc - bcc \\ - ccc - bdc \\ - dcc + cdc = -bcd. \end{aligned}$$

But, once terms of opposite signs have been removed, the truth of this equality is obvious.

Therefore having put c equal to the root a, it is indeed [shown to be] equal.

Likewise if d is put equal to the root a, then changing a into d we get
$$\begin{aligned} ddd + bdd - bcd \\ - cdd - bdd \\ - ddd + cdd = -bcd. \end{aligned}$$

And the truth of this equality is similarly obvious.

Therefore having put d equal to the root a, it is indeed [shown to be] equal.

Thus c & d are roots equal to the sought root a, as was stated in the proposition.

And in the following Lemma it is demonstrated that there is no other root besides **55** c and d equal to the root a of the equation.

LEMMA

Suppose another root of the equation equal to the root a and not equal to the roots c and d could be given; let it be b (or anything else).

Then putting $b = a$ it will be true that
$$\begin{array}{l} bbb + bbb - bcb \\ \quad - cbb - bdb \\ \quad - dbb + cdb = -bcd. \end{array}$$

And on rearranging the terms $+2bbb - 2bbd = +2cbb - 2cbd$.

i.e. $+bbb - bbd = +cbb - bcd$.

Therefore $\dfrac{+bb - bd}{b} = \dfrac{+bb - bd}{c}$

Therefore $b = c$ which is contrary to the hypothesis.

Alternatively $+2bbb - 2bbc = +2dbb - 2dbc$.

i.e. $+bbb - bbc = +2dbb - dbc$.

Therefore $\dfrac{+bb - bc}{b} = \dfrac{+bb - bc}{d}$

Therefore $b = d$ which is also contrary to the hypothesis.

Therefore it is not true that $b = a$ as was assumed. And by precisely the same reasoning the same thing can be concluded about any other root apart from c and d.

COROLLARY

It is clear by simple inspection that the pairs of equations in the two theorems given above can be joined.

For
$$\begin{array}{l} +aaa + baa + bca \\ \quad + caa - bda \\ \quad - daa - cda = +bcd = \end{array} \begin{array}{l} + bca - baa \\ \quad + bda + baa \\ \quad - cda + daa - aaa. \end{array}\ \text{[8]}$$

Moreover, the relationships between the roots is known from the theorems; firstly that $a = b$, and secondly that $a = c$ or d, which is the important point to notice.

PROPOSITION 5

b and c and d are the explicatory roots of the equation
$$\begin{array}{l} aaa - baa + bca \\ \quad - caa + bda \\ \quad - daa + cda = +bcd, \end{array}\quad \text{equal to the sought root } a.$$

56 For if b is put equal to the root a of the equation
$$\begin{array}{l} aaa - baa + bca \\ \quad - caa + bda \\ \quad - daa + cda = +bcd, \end{array}$$

then, changing a into b we get
$$\begin{array}{l} bbb - bbb + bcb \\ \quad - cbb + bdb \\ \quad - dbb + cdb = +bcd \end{array}$$

But, once terms of opposite signs have been removed, this equality is obvious.

Therefore having put b equal to the root a, it is indeed [shown to be] equal.

Likewise, if c is put equal to the root a, then, changing a into c we get
$$ccc - bcc + bcc$$
$$- ccc + bdc$$
$$- dcc + cdc = +bcd.$$

And this equality is obvious, once terms of opposite signs have been removed.

Therefore having put c equal to the root a, it is indeed [shown to be] equal.

Likewise if d is put equal to the root a, then, changing a into d we get
$$ddd - bdd + bcd$$
$$- cdd + bdd$$
$$- ddd + cdd = +bcd.$$

And this equality is obvious, once terms of opposite signs have been removed.

Therefore having put d equal to the root a, it too is [shown to be] equal.

Thus b, c and d are roots equal to the sought root a, as was stated in the proposition.

And in the following Lemma it is demonstrated that there is no other root besides b and c and d equal to the root a.

LEMMA

Suppose another root of the equation equal to the root a and not equal to the roots b and c and d could be given; let it be f (or anything else).

Then putting $f = a$ it will be true that
$$fff - bff + bcf$$
$$- cff + bdf$$
$$- dff + cdf = +bcd.$$

And on rearranging the terms $fff - cff + cdf - dff = +bff - bcf + bcd - bdf$.

Therefore $$\left.\frac{ff - cf + cd - df}{f}\right| = \left.\frac{+ff - cf + cd - df}{b}\right|$$

Therefore $f = b$ which is contrary to the hypothesis.

Or, taking it in a different order: $fff - bff + bdf - dff = cff - cbf + bcd - cdf$

Therefore $$\left.\frac{ff - bf + bd - df}{f}\right| = \left.\frac{ff - bf + bd - df}{c}\right|$$

Therefore $f = c$ which is also [against] the hypothesis.

Or, taking it in yet a different order: $fff - bff + bcf - cff = dff - dbf + dbc - dcf$.

Therefore $$\left.\frac{ff - bf + bc - cf}{f}\right| = \left.\frac{ff - bf + bc - cf}{d}\right|$$

57

Therefore $f = d$ which is also contrary to the hypothesis.

Therefore it is not true that $f = a$ as was assumed. And by similar reasoning the same may be concluded about any other root apart from b, c and d.

Reduced Equations

PROPOSITION 6

b and c are the roots of the equation
$$\begin{aligned}aaa - bba\\ - bca\\ - cca = -\ bbc\\ - bcc\end{aligned}$$
equal to the sought

root a.

For if we put $b = a$, and in the proposed equation if a is changed into b, then we

get $\quad \begin{matrix} bbb - bbb \\ -bbc \\ -bcc = -bbc \\ -bcc \end{matrix}$

Or if we put $c = a$ and if a is changed into c, then we get $\quad \begin{matrix} ccc - bbc \\ -bcc \\ -ccc = -bbc \\ -bcc \end{matrix}$

But these equalities are obvious, once terms of opposite signs have been removed.

Thus in the proposed equation the sought root $a = b$ or c, as was stated in the proposition.

PROPOSITION 7

$b + c$ is the root of the equation $\quad \begin{matrix} aaa - bba \\ -bca \\ -cca = +bbc \\ +bcc \end{matrix}$ equal to the sought root a.

58 For if we put $b + c = a$ and in the proposed equation if a is changed into $b + c$,

then we get $\quad \begin{matrix} +bbb \quad -bbb \\ +3bbc \quad -bbc \\ +3bcc \quad -bbc \\ +ccc \quad -bcc \\ \quad\quad -bcc \\ \quad -ccc = +bbc \\ \quad\quad +bcc \end{matrix}$

But this equality is obvious, once terms of opposite signs have been removed; that

is $\quad \begin{matrix} +bbc \\ +bcc = +bbc \\ +bcc. \end{matrix}$

Thus, having put $b + c$ equal to the root a, it is indeed [shown to be] equal, as was stated in the proposition.

COROLLARY

From this it is obvious that this equation can be joined to the immediately previous one.

For $\quad \begin{matrix} aaa - bba \\ -bca \\ -cca = +bbc \\ \quad\quad +bcc = +bba \\ \quad\quad\quad +bca \\ \quad\quad\quad +cca - aaa. \end{matrix}$

And it is enough to note that in the first equation $a = b + c$, and in the second $a = b$ or c.

PROPOSITION 8

b and c are the roots of the equation $\quad \begin{matrix} aaa - bbaa = -\dfrac{bbcc}{b+c} \\ \dfrac{-bcaa}{} \\ \dfrac{-ccaa}{b+c} \end{matrix}$ equal to the sought

root a.

For if we put $b = a$, then we get
$$\frac{+bbbb - bbbb}{b+c} \frac{= -bbcc}{b+c}$$
$$\frac{+cbbb - cbbb}{b+c - ccbb}$$
$$\frac{}{b+c}$$

Or putting $c = a$, we get
$$\frac{+bccc - bbcc}{b+c} \frac{= -bbcc}{b+c}$$
$$\frac{+cccc - bccc}{b+c - cccc}$$
$$\frac{}{b+c}$$

But these equalities are obvious.

Thus in the proposed equation, the root $a = b$ or c, as was stated in the proposition.

PROPOSITION 9

59

$\frac{bc}{b+c}$ is the root of the equation
$$\frac{aaa + bbaa}{+ bcaa} \frac{}{}$$
$$\frac{+ ccaa}{b+c} = \frac{+bbcc}{b+c},$$
equal to the sought root a.

For (by Section 3, Problem 5) the binomial equation proposed here is reduced from its trinomial equation, by putting $\frac{bc}{b+c} = d$, and changing the one into the other.

But (by Proposition 3 of this section) that trinomial has the root $a = d$.

Thus this binomial equation has the root $a = \frac{bc}{b+c}$ as was stated in the proposition.

COROLLARY

From this it is obvious that this equation can be joined to the immediately previous one.

For
$$\frac{aaa + bbaa}{+ bcaa} \frac{}{}$$
$$\frac{+ ccaa}{b+c} = \frac{+bbcc}{b+c} = \frac{+bbaa}{+bcaa}$$
$$\frac{+ccaa - aaaa}{b+c}$$

And it is enough to note that in the first equation $a = \frac{bc}{b+c}$, and in the second $a = b$, or c.

PROPOSITION 10

$c - b$ is the root of the equation $aaa + 3baa + 3bba = +ccc - bbb$ equal to the sought root a.

For if we put $c - b$ equal to the root a of the equation $aaa + 3baa + 3bba = +ccc - bbb$ then, changing a into $c - b$, we get

$$\left. \begin{array}{l} +ccc - 3bcc + 3bbc - \ bbb = \quad aaa \\ \text{And} + 3bcc - 6bbc + 3bbb = +3baa \\ \text{And} \qquad\quad + 3bbc - 3bbb = +3bba \end{array} \right\} = +ccc - bbb.$$

But this equality is obvious, once terms of opposite signs have been removed.

Thus having put $c - b$ equal to the root a, it is indeed [shown to be] equal, as was stated in the proposition.

PROPOSITION 11

60 $c + b$ is the root of the equation $aaa - 3baa + 3bba = +ccc + bbb$, equal to the sought root a.

For if we put $c + b$. equal to the root a of the equation $aaa - 3baa + 3bba = +ccc + bbb$, then changing a into $c + b$ we get

$$\left. \begin{array}{l} +ccc + 3bcc + 3bbc + bbb = +aaa \\ \text{And} - 3bcc - 6bbc - 3bbb = -3baa \\ \text{And} + 3bbc + 3bbb = +3bba \end{array} \right\} = +ccc + bbb$$

But this equality is obvious, once terms of opposite signs have been removed.
Thus, having put $c + b$ equal to the root a, it is indeed [shown to be] equal, as was stated in the proposition.

PROPOSITION 12

$b - c$ is the root of the equation $aaa - 3baa + 3bba = +bbb - ccc$, equal to the sought root a.

For if we put $b - c$ equal to the root a of the equation $aaa - 3baa + 3bba = +bbb - ccc$, then changing a into $b - c$ we get

$$\left. \begin{array}{l} -ccc + 3bcc - 3bbc + bbb = +aaa \\ \text{And} - 3bcc + 6bbc - 3bbb = -3baa \\ \text{And} - 3bbc + 3bbb = +3bba \end{array} \right\} = +bbb - ccc.$$

But this equality is obvious, once terms of opposite signs have been removed.
Thus, having put $b - c$ equal to the root a, it is indeed [shown to be] equal, as was stated in the proposition.

PROPOSITION 13

$2b$ is the root of the equation $aaa - 3baa + 3bba = +2bbb$ equal to the sought root a.

For if we put $2b$ is equal to the root a of the equation above, then changing a into $2b$ we get $8bbb - 12bbb + 6bbb = +2bbb$

But this equality is obvious in and of itself.

Thus, having put $2b$ equal to the root a, it is indeed [shown to be] equal, as was stated in the proposition.

Reduced Equations

PROPOSITION 14   61

$c - b$ is the root of the equation $aaa + 3bca = +ccc - bbb$, equal to the sought root a.

For if we put $c - b$ equal to the root a of the equation $aaa + 3bca = +ccc - bbb$, then, changing a into $c - b$ we get

$$\left.\begin{array}{l} ccc - 3bcc + 3bbc - bbb = +aaa \\ \text{And} \qquad\quad + 3bcc - 3bbc = +3bca \end{array}\right\} = +ccc - bbb$$

But this equality is obvious, once terms of opposite signs have been removed.
Thus, having put $c - b$ equal to the root a, it is indeed [shown to be] equal, as was stated in the proposition.

PROPOSITION 15

$c + b$ is the root of the equation $aaa - 3bca = +ccc + bbb$, equal to the sought root a.

For if we put $c + b$ equal to the root a of the equation $aaa - 3bca = +ccc + bbb$, then, changing a into $c + b$ we get

$$\left.\begin{array}{l} ccc + 3bcc + 3bbc + bbb = +aaa \\ \text{And} \qquad\quad - 3bcc - 3bbc = -3bca \end{array}\right\} = +ccc + bbb$$

But this equality is obvious, once terms of opposite signs have been removed.
Thus having put $c + b$, equal to the root a, it is indeed [shown to be] equal, as was stated in the proposition.

PROPOSITION 16

$b - c$ is the root of the equation $aaa + 3bca = -ccc + bbb$, equal to the sought root a.

For if we put $b - c$ equal to the root a of the equation $aaa + 3bca = -ccc + bbb$, then, changing a into $b - c$ we get

$$\left.\begin{array}{l} bbb - 3cbb + 3ccb - ccc = +aaa \\ \text{And} \qquad\quad + 3cbb - 3ccb = +3bca \end{array}\right\} = -ccc + bbb$$

But this equality is obvious, once terms of opposite signs have been removed.
Thus having put $b - c$ equal to the root a, it is indeed [shown to be] equal, as was stated in the proposition.

PROPOSITION 17   62

$2b$ is the root of the equation $aaa - 3bba = +2bbb$, equal to the sought root a.

For if we put $2b$ equal to the root a of the equation $aaa - 3bba = +2bbb$, then, changing a into $2b$ we get $8bbb - 6bbb = +2bbb$

But this equality is obvious, once terms of opposite signs have been removed.

Thus having put $2b$ equal to the root a, it is indeed [shown to be] equal, as was stated in the proposition.

Reciprocal Equations

PROPOSITION 18

b is the root of the equation $aaa - baa + cda = +bcd$, equal to the sought root a.

For if we put b equal to the root a of the equation $aaa - baa + cda = +bcd$, then, changing a into b we get $bbb - bbb + cdb = +bcd$

But this equality is obvious in and of itself.

Thus having put b equal to the root a, it is indeed [shown to be] equal, as was stated in the proposition.

PROPOSITION 19

c is the root of the equation $aaa + baa - caa = +bcc$, equal to the sought root a.

For if we put c equal to the root a of the equation $aaa + baa - cca = +bcc$, then, changing a into c we get $ccc + bcc - ccc = +bcc$

But this equality is obvious in and of itself.

Thus having put c equal to the root a, it is indeed [shown to be] equal, as was stated in the proposition.

PROPOSITION 20

b and c are the roots of the equation $aaa - baa - cca = -bcc$, equal to the sought root a.

For if we put b equal to the root a of the equation $aaa - baa - cca = -bcc$, then, changing a into b we get $bbb - bbb - ccb = -bcc$

63   But this equality is obvious in and of itself.

Thus, having put b equal to the root a, it is indeed [shown to be] equal.

Likewise, if we put c equal to the root a, then, changing a into c we get $ccc - bcc - ccc = -bcc$

And this equality is also obvious in and of itself.

Thus, having put c equal to the root a, it is also [shown to be] equal.

Thus b and c are roots equal to the sought root a, as was stated in the proposition.

PROPOSITION 21

f is the root of the equation

$$aaaa + baaa + bcaa \\ + caaa + bdaa \\ + daaa + cdaa + bcda \\ - faaa - bfaa - bcfa \\ - cfaa - bdfa \\ - dfaa - cdfa = +bcdf,$$

equal to the sought root a.

For if we put f equal to the root a of the proposed equation, then, changing a into f we get

$$\begin{aligned}
ffff &+ bfff &+ bcff \\
&+ cfff &+ bdff \\
&+ dfff &+ cdff &+ bcdf \\
&- ffff &- bfff &- bcff \\
&&- cfff &- bdff \\
&&- dfff &- cdff = +bcdf
\end{aligned}$$

But this equality is obvious, once terms of opposite signs have been removed.

Thus, having put f equal to the root a, it is indeed [shown to be] equal, as was stated in the proposition.

And in the following Lemma it is demonstrated that there is no other root besides f equal to the root a of the equation.

LEMMA

Suppose another root of the equation equal to the root a and not equal to f could be given; let it be b or c or d (or anything else).

Then putting $b = a$, it will be true that

$$\begin{aligned}
bbbb &+ bbbb &+ bbbc \\
&+ cbbb &+ bbbd \\
&+ dbbb &+ bbcd &+ bbcd \\
&- fbbb &- bbbf &- bbcf \\
&&- bbcf &- bbdf \\
&&- bbdf &- bcdf = +bcdf
\end{aligned}$$

Therefore* $+2bbbb + 2bbbc + 2bbbd + 2bbcd = +2bbbf + 2bbcf + 2bbdf + 2bcdf$ [9]

i.e., $+bbbb + bbbc + bbbd + bbcd = +bbbf + bbcf + bbdf + bcdf$ **64**

Therefore $\dfrac{+bbb + bbc + bbd + bcd}{f} = \dfrac{+bbb + bbc + bbd + bcd}{b}$

Therefore $c = f$, which is contrary to the hypothesis of the Lemma.

Therefore it is not true that $b = a$, as was assumed; and by [similar] reasoning the same may also be concluded concerning c or d, or any other root apart from f.

PROPOSITION 22

b and c and d, are the roots of the equation

$$\begin{aligned}
aaa &- baaa &+ bcaa \\
&- caaa &+ bdaa \\
&- daaa &+ cdaa &- bcda \\
&+ faaa &- bfaa &+ bcfa \\
&&- cfaa &+ bdfa \\
&&- dfaa &+ cdfa = +bcdf,
\end{aligned}$$

equal to the sought root a.

For if we put b equal to the root a of the proposed equation, then, changing a into b we get

$$\begin{aligned}
bbbb &- bbbb &+ bcbb \\
&- cbbb &+ bdbb \\
&- dbbb &+ cdbb &- bcdb \\
&+ fbbb &- bfbb &+ bcfb \\
&&- cfbb &+ bdbb \\
&&- dfbb &+ cdfb = +bcdf
\end{aligned}$$

---

*Because of the length of the expressions, the original is printed with the right-hand side over the left-hand side and a vertical mark of equality between them.

But this equality is obvious, once terms of opposite signs have been removed.

Thus, having put b equal to the root a, it is indeed [shown to be] equal.

Likewise, if we put c equal to the root a, then, changing a into c we get

$$
\begin{aligned}
cccc - bccc &+ bccc \\
- cccc &+ bdcc \\
- dccc &+ cdcc - bcdc \\
+ fccc &- bfcc + bcfc \\
&- cfcc + bdfc \\
&- dfcc + cdfc = +bcdf
\end{aligned}
$$

But this equality is obvious, once terms of opposite signs have been removed.

Thus, having put c equal to the root a, it is indeed [shown to be] equal.

Likewise, if we put d equal to the root a, then, changing a into d we get

$$
\begin{aligned}
dddd - bddd &+ bcdd \\
- cddd &+ bddd \\
- dddd &+ cddd - bcdd \\
+ fddd &- bfdd + bcfd \\
&- cfdd + bdfd \\
&- dfdd + cdfd = +bcdf
\end{aligned}
$$

65 But this equality is obvious, once terms of opposite signs have been removed.

Thus, having put c equal to the root a, it is indeed [shown to be] equal.

Thus b, c and d are roots equal to the sought root a, as was stated in the proposition.

And in the following Lemma it is demonstrated that there is no other root besides b, c and d equal to the root a of the equation.

LEMMA

Suppose another root of the equation equal to the root a and not equal to b, c or d could be given; let it be f (or anything else).

Then putting $f = a$, it will be true that

$$
\begin{aligned}
ffff - bfff &+ bcff \\
- cfff &+ bdff \\
- dfff &+ cdff - bcdf \\
+ ffff &- bfff + bcff \\
&- cfff + bdff \\
&- dfff + cdff = +bcdf
\end{aligned}
$$

Therefore $+2ffff - 2cfff + 2cdff - 2dfff = +2bfff - 2bcff + 2bcdf - 2bdff$

i.e., $+ffff - cfff + cdff - dfff = +bfff - bcff + bcdf - bdff$

$$
\frac{+fff - cff + dcf - dff}{f} = \frac{+fff - cff + cdf - dff}{b}
$$

Therefore $f = b$, which is contrary to the hypothesis of the Lemma.

Therefore it is not true that $f = a$, as was assumed. And by similar reasoning the same may be demonstrated concerning any other root.

PROPOSITION 23

b and c are the roots of the equation

$$\begin{aligned}
aaaa &- baaa + bcaa \\
&- caaa - bdaa \\
&+ daaa - bfaa + bcda \\
&+ faaa - cdaa + bcfa \\
&- cfaa - bdfa \\
&+ dfaa - cdfa = -bcdf
\end{aligned}$$

equal to the sought root a.

For if we put b equal to the root a of the proposed equation, then, changing a into b we get

66

$$\begin{aligned}
bbbb &- bbbb + bcbb \\
&- cbbb - bdbb \\
&+ dbbb - bfbb + bcdb \\
&+ fbbb - cdbb + bcfb \\
&- cfbb - bdfb \\
&+ dfbb - cdfb = -bcdf
\end{aligned}$$

But this equality is obvious, once terms of opposite signs have been removed.

Thus, having put b equal to the root a, it is indeed [shown to be] equal.

Likewise if we put c, equal to the root a of the proposed equation, then, changing a into c we get

$$\begin{aligned}
+cccc &- bccc + bccc \\
&- cccc - bdcc \\
&+ dccc - bfcc + bcdc \\
&+ fccc - cdcc + bcfc \\
&- cfcc - bdfc \\
&+ dfcc - cdfc = -bcdf
\end{aligned}$$

But this equality is obvious, once terms of opposite signs have been removed.

Thus, having put c equal to the root a, it also is [shown to be] equal.

Thus b and c are roots equal to the sought root a, as was stated in the proposition. And in the following Lemma it is demonstrated that there is no other root besides b and c equal to the root a of the equation.

LEMMA

Suppose another root of the equation equal to the root a and not equal to b or c could be given; let it first be d or f.

Then putting $d = a$, it will be true that

$$\begin{aligned}
dddd &- bddd + bcdd \\
&- cddd - bddd \\
&+ dddd - bfdd + bcdd \\
&+ fddd - cddd + bcfd \\
&- cfdd - bfdd \\
&+ fddd - cfdd = -bcdf
\end{aligned}$$

Therefore $+2dddd - 2cddd + 2fddd - 2cfdd = +2bddd - 2bcdd + 2bfdd - 2bcdf$

i.e., $+bddd - bcdd + bfdd - bcdf = +dddd - cddd + fddd - cfdd$

Therefore $\left. \dfrac{ddd - cdd + fdd - cfd}{d} \right| = \left. \dfrac{+ddd - cdd + fdd - cfd}{b} \right|$

Therefore $d = b$, which is contrary to the hypothesis of the Lemma.

We reach a similar contradiction with the conclusion $d = c$, if the sixteen terms of the equation are similarly arranged with respect to c.

Therefore it is not true that $d = a$, as was assumed; and by similar reasoning the same may be stated concerning f or any other root apart from b and c.

67   PROPOSITION 24

b, c, d and f are roots of the equation

$$aaaa - baaa + bcaa$$
$$- caaa + bdaa$$
$$- daaa + bfaa - bcda$$
$$- faaa + cdaa - bcfa$$
$$+ cfaa - bdfa$$
$$+ dfaa - cdfa = -bcdf$$

equal to the sought root a.

For if we put b equal to the root a of the proposed equation, then, changing a into b we get

$$bbbb - bbbb + bcbb$$
$$- cbbb + bdbb$$
$$- dbbb + bfbb - bcdb$$
$$- fbbb + cdbb - bcfb$$
$$+ cfbb - bdfb$$
$$+ dfbb - cdfb = -bcdf.$$

But this equality is obvious, once terms of opposite signs have been removed.

Thus, having put b equal to the root a, it is indeed [shown to be] equal.

Likewise, if we put c equal to the root a of the proposed equation, then, changing a into c we get

$$cccc - bccc + bccc$$
$$- cccc + bdcc$$
$$- dccc + bfcc - bcdc$$
$$- fccc + cdcc - bcfc$$
$$+ cfcc - bdfc$$
$$+ dfcc - cdfc = -bcdf$$

And this equality is also obvious, once terms of opposite signs have been removed.

Thus, having put c equal to the root a, it is also [shown to be] equal.

Likewise, if we put d or f equal to the root a, similar equalities follow from the substitution; and thus we must similarly conclude that these too are roots equal to the root a.

Thus b, c, d and f are roots equal to the sought root a, as was stated in the proposition.

In the following Lemma it is demonstrated that there is no other root besides b, c, d and f equal to the root a of the equation.

LEMMA

Suppose another root of the equation equal to the root a and not equal to b, c, d or f could be given; let it be g (or anything else).

Then putting $g = a$, it will be true that

$$\begin{aligned}
gggg &- bggg + bcgg \\
&- cggg + bdgg \\
&- dggg + cdgg - bcdg \\
&- fggg + bfgg - bcfg \\
&+ cfgg - bdfg \\
&+ dfgg - cdfg = -bcdf
\end{aligned}$$

68

Therefore $gggg - cggg + cdgg - dggg + cfgg - fggg + dfgg - cdfg = bggg - bcgg + bcdg - bdgg + bcfg - bfgg + bdfg - bcdf$

Therefore

$$\frac{ggg - cgg + cdg - dgg + cfg - fgg + dfg - cdf}{g} =$$
$$\frac{ggg - cgg + cdg - dgg + cfg - fgg + dfg - cdf}{b}$$

Therefore $g = b$, which is contrary to the hypothesis of the Lemma.

We reach a similar contradiction with the conclusions $g = c$, $g = d$ and $g = f$, if (following similar reasoning) the sixteen terms of the equation are appropriately arranged with respect to c, d or f. But let the proof here for b suffice as a model of how the falsity of the subsitution should be demonstrated in the other cases.

Thus it is not true that $g = a$, as was assumed; and by identical reasoning the same may be stated concerning any other root.

Reduced Equations

PROPOSITION 25

b, c, and d, are the roots of the equation
$$\begin{aligned}
aaaa &- bbaa + bbca \\
&- ccaa + bbda \\
&- ddaa + bcca \\
&- bcaa + ccda \\
&- bdaa + bdda \\
&- cdaa + cdda = + bbcd \\
&\quad\quad\; + 2bcda \;\; + bccd \\
&\quad\quad\quad\quad\quad + bcdd,
\end{aligned}$$
equal

to the sought root a.

For if we put b equal to the root a of the proposed equation, then, changing a, into b we get

$$\begin{aligned}
bbbb &- bbbb + bbcb \\
&- ccbb + bbdb \\
&- ddbb + bccb \\
&- bcbb + ccdb \\
&- bdbb + bddb \\
&- cdbb + cddb \\
&\quad\quad + 2bcdb = + bbcd \\
&\quad\quad\quad\quad\quad + bccd \\
&\quad\quad\quad\quad\quad + bcdd
\end{aligned}$$

69

But this equality is obvious, once terms of opposite signs have been removed.

Thus, having put b equal to the root a, it is indeed [shown to be] equal.

Likewise, if we put c equal to the root a of the proposed equation, then, changing a into c we get

$$
\begin{aligned}
cccc &- bbcc + bbcc \\
&- cccc + bbdc \\
&- ddcc + bccc \\
&- bccc + ccdc \\
&- bdcc + bddc \\
&- cdcc + cddc \\
&+ 2bcdc = + bbcd \\
& + bccd \\
& + bcdd
\end{aligned}
$$

But this equality is obvious, once terms of opposite signs have been removed.

Thus having put c equal to the root a, it is indeed [shown to be] equal.

Likewise, if we put d equal to the root a of the proposed equation, then, changing a into d we get

$$
\begin{aligned}
dddd &- bbdd + bbcd \\
&- ccdd + bbdd \\
&- dddd + bccd \\
&- bcdd + ccdd \\
&- bddd + bddd \\
&- cddd + cddd \\
&+ 2bcdd = + bbcd \\
& + bccd \\
& + bcdd
\end{aligned}
$$

And this equality is also obvious, once terms of opposite signs have been removed.

Thus having put d equal to the root a, it is also [shown to be] equal.

Thus b, c and d are roots equal to the sought root a, as was stated in the proposition.

PROPOSITION 26

b, c and d are the roots of the equation.

$$
\begin{aligned}
aaaa &- bbaaa + bbcca \\
&- ccaaa + bbdda \\
&- ddaaa + ccdda \\
&- bcaaa + bcdda \\
&- bdaaa + bccda \\
&\dfrac{- cdaaa + bbcda}{b+c+d \;\; b+c+d} = \dfrac{+ bbccd}{} \\
&\phantom{\dfrac{- cdaaa}{b}} + bbcdd \\
&\phantom{\dfrac{- cdaaa}{b}} + bccdd \\
&\phantom{\dfrac{- cdaaa}{b}} \overline{b+c+d,}
\end{aligned}
$$

equal to the sought root a.

70 For if we put b equal to the root a of the proposed equation, then, changing a into b and expressing the power [i.e. of b alone, $bbbb$] in terms of the common divisor, we get

$$
\begin{aligned}
&+ bbbbb - bbbbb + bbccb \\
&+ cbbbb - ccbbb + bbcdb \\
&\dfrac{+ dbbbb - ddbbb + ccddb}{b+c+d} - bcbbb + bcddb \\
&\phantom{\dfrac{+ dbbbb}{b}} - bdbbb + bccdb \\
&\phantom{\dfrac{+ dbbbb}{b}} \dfrac{- cdbbb + bbcdb}{b+c+d \;\; b+c+d} = \dfrac{+ bbccd}{} \\
&\phantom{\dfrac{+ dbbbb}{bbb}} + bbcdd \\
&\phantom{\dfrac{+ dbbbb}{bbb}} + bccdd \\
&\phantom{\dfrac{+ dbbbb}{bbb}} \overline{b+c+d}
\end{aligned}
$$

But this equality is obvious, once terms of opposite signs have been removed.

Thus, having put b equal to the root a, it is indeed [shown to be] equal.

Likewise, if we put c or d equal to the root a, similar equalities follow from the substitution; and thus we must similarly conclude that these too are roots equal to the root a.

Thus b, c and d are roots equal to the sought root a, as was stated in the proposition.

PROPOSITION 27

b, c and d are the roots of the equation

$$\frac{\begin{matrix} aaaa - bbcaaa \\ -bbdaaa \\ -bccaaa \\ -bddaaa \\ -ccdaaa \\ -cddaaa \\ -2bcdaaa \end{matrix}}{bc + bd + cd} \frac{\begin{matrix} +bbccaa \\ +bbddaa \\ +ccddaa \\ +bcddaa \\ +bccdaa \\ +bbcdaa \end{matrix}}{bc + bd + cd} \quad \frac{= +bbccdd}{bc + bd + cd} \qquad \text{equal to the sought}$$

root a.

For if we put b equal to the root a of the proposed equation, then, changing a into b and expressing the power in terms of the common divisor, we get

$$\frac{\begin{matrix} + bcbbbb \\ + bdbbbb \\ + cdbbbb \end{matrix}}{bc + bd + cd} \frac{\begin{matrix} - bbcbbb \\ - bbdbbb \\ - bccbbb \\ - bddbbb \\ - ccdbbb \\ - cddbbb \\ - 2bcdbbb \end{matrix}}{bc + bd + cd} \frac{\begin{matrix} + bbccbb \\ + bbddbb \\ + ccddbb \\ + bcddbb \\ + bccdbb \\ + bbcdbb \end{matrix}}{bc + bd + cd} \quad \frac{= + bbccdd}{bc + bd + cd}$$

But this equality is obvious, once terms of opposite signs have been removed.   **71**

Therefore, having put b equal to the root a, it is indeed [shown to be] equal.

Likewise, if we put c or d equal to the root a, similar equalities follow from the substitution; and thus we must similarly conclude that these too are roots equal to the root a.

Thus b, c and d are roots equal to the sought root a, as was stated in the proposition.

PROPOSITION 28

If it should be the case that

$$\begin{matrix} aaaa - bbaa - bbca \\ - ccaa - bbda \\ - ddaa - bcca \\ - bcaa - ccda \\ - bdaa - bdda \\ - cdaa - cdaa \\ - 2bcda = + bbcd \\ + bccd \\ + bcdd \end{matrix}$$

then $b + c + d$ is the root equal to the sought root a.

For (by Sect.3, Problem 12) the trinomial equation proposed here is deduced from its quadrinomial equation by putting $b + c + d = f$.

But (by Propos. 21. of this section) that quadrinomial has the root $a = f$.

Thus this trinomial has the root $a = b + c + d$, as was stated in the proposition.

PROPOSITION 29

$\frac{bc+bd+cd}{b+c+d}$ is the root of the equation

$$
\begin{array}{ll}
aaaa + bbaaa & - bbcca \\
+ ccaaa & - bbdda \\
+ ddaaa & - ccdda \\
+ bcaaa & - bcdda \\
+ bdaaa & - bccda \\
+ cdaaa & - bbcda = + bbccd \\
\hline
b+c+d\ \overline{b+c+d} & + bbcdd \\
& + bccdd \\
& \overline{b+c+d}
\end{array}
$$

equal to the sought root a.

For (by Sect. 3, Probl. 13) the trinomial equation proposed here is deduced from **72** its quadrinomial equation by putting $\frac{bc+bd+cd}{b+c+d} = f$.

But (by Prob. 21 of this section) that quadrinomial has the root $a = f$.

Thus this trinomial has the root $a = \frac{bc+bd+cd}{b+c+d}$ as was stated in the proposition.

PROPOSITION 30

$\frac{bcd}{bc+bd+cd}$ is the root of the equation

$$
\begin{array}{lll}
aaaa + bbcaaa & & \\
+ bbdaaa & + bbccaa & \\
+ bccaaa & + bbddaa & \\
+ bddaaa & + ccddaa & \\
+ ccdaaa & + bcddaa & \text{equal to the sought} \\
+ cddaaa & + bccdaa & \\
+ 2bcdaaa & + bbcdaa & = + bbccdd \\
\hline
bc+bd+cd\ \overline{bc+bd+cd} & & \overline{bc+bd+cd}
\end{array}
$$

root a.

For (by Sect 3., Probl. 14) the trinomial equation proposed here is deduced from its quadrinomial equation by putting $\frac{bcd}{bc+bd+cd} = f$.

But (by Probl. 21. of this Sect.) that quadrinomial has the root $a = f$.

Consequently this trinomial has the root $a = \frac{bcd}{bc+bd+cd}$ as was stated in the proposition.

PROPOSITION 31

b and c are the roots of the equation

$$
\begin{array}{ll}
aaaa + bdaa & + bbca \\
+ cdaa & + bcca \\
- bbaa & + bdda \\
- bcaa & + cdda \\
- ccaa & - bbda \\
- ddaa & - ccda \\
& - 2bcda = - bbcd \\
& - bccd \\
& + bcdd,
\end{array}
$$
equal

to the sought root a.

For if we put b equal to the root a of the proposed equation, then, changing a **73**
into b we get

$$\begin{array}{l} bbbb + bdbb + bbcb \\ \quad + cdbb + bccb \\ \quad - bbbb + bddb \\ \quad - bcbb + cddb \\ \quad - ccbb - bbdb \\ \quad - ddbb - ccdb \\ \qquad - 2bcdb = - bbcd \\ \qquad\qquad\qquad\quad - bccd \\ \qquad\qquad\qquad\quad + bcdd \end{array}$$

But this equality is obvious, once terms of opposite signs have been removed.

Thus, having put b equal to the root a, it is indeed [shown to be] equal.

Likewise, if we put c equal to the root a, a similar equality follows from the substitution; and thus we must similarly conclude that that too is a root equal to the root a.

Therefore b and c are roots equal to the sought root a, as was stated in the proposition.

PROPOSITION 32

b and c are the roots of the equation

$$\frac{\begin{array}{l} aaaa - bbaaa + bbcca \\ \quad - bcaaa + bbdda \\ \quad - ccaaa + bcdda \\ \quad - ddaaa + ccdda \\ \quad + bdaaa - bbcda \\ \quad + cdaaa - bccda \end{array}}{b + c - d \;\; b + c - d} = \frac{\begin{array}{l} - bbccd \\ + bbcdd \\ + bccdd \end{array}}{b + c - d} \quad \text{equal}$$

to the sought root a.

For if we put b equal to the root a of the proposed equation, then, changing a into b and expressing the substituted power in terms of the common divisor, we get

$$\frac{\begin{array}{l} + bbbbb - bbbbb + bbccb \\ \quad + cbbbb - bcbbb + bbddb \\ \quad - dbbbb - ccbbb + bcddb \\ \quad - ddbbb + ccddb \\ \quad + bdbbb - bbcdb \\ \quad + cdbbb - bccdb \end{array}}{b + c - d \;\; b + c - d} = \frac{\begin{array}{l} - bbccd \\ + bbcdd \\ + bccdd \end{array}}{b + c - d}$$

But this equality is obvious, once terms of opposite signs have been removed.

Thus, having put b equal to the root a, it is indeed [shown to be] equal.

Likewise, if we put c equal to the root a, a similar equality follows from the **74**
substitution; and thus we must similarly conclude that that too is a root equal to
the root a.

Thus b and c are roots equal to the sought root a, as was stated in the proposition.

PROPOSITION 33

b and c are the roots of the equation

$$\frac{\begin{aligned}aaaa + bbaaa \quad &- bbcca\\ + bcaaa \quad &- bbdda\\ + ccaaa \quad &- bcdda\\ + ddaaa \quad &- ccdda\\ - bdaaa \quad &+ bbcda\\ - cdaaa \quad &+ bccda\end{aligned}}{d-b-cd-b-c} = \frac{\begin{aligned}&- bbcdd\\ &- bccdd\\ &+ bbccd\end{aligned}}{d-b-c}$$

equal to the sought root a.

For if we put b equal to the root a of the proposed equation, then, changing a into b and expressing the substituted power in terms of the common divisor, we get

$$\frac{\begin{aligned}+ dbbbb \quad + bbbbb \quad &- bbccb\\ - bbbbb \quad + bcbbb \quad &- bbddb\\ - cbbbb \quad + ccbbb \quad &- bcddb\\ \overline{d-b-c} \quad + ddbbb \quad &- ccddb\\ - bdbbb \quad &+ bbcdb\\ - cdbbb \quad &+ bccdb\end{aligned}}{d-b-cd-b-c} = \frac{\begin{aligned}&- bbcdd\\ &- bccdd\\ &- bbccd\end{aligned}}{d-b-c}$$

But this equality is obvious, once terms of opposite signs have been removed.

Thus having put b equal to the root a, it is indeed [shown to be] equal.

Likewise, if we put c equal to the root a, a similar equality follows from the substitution; and thus we must similarly conclude that that too is a root equal to the root a.

Thus b and c are roots equal to the sought root a, as was stated in the proposition.

PROPOSITION 34

b and c are the roots of the equation

$$\frac{\begin{aligned}aaaa + bbcaaa \quad &\\ + bccaaa \quad &- bbccaa\\ + bddaaa \quad &- bbddaa\\ + cddaaa \quad &- bcddaa\\ - bbdaaa \quad &- ccddaa\\ - ccdaaa \quad &+ bccdaa\\ - 2.bcdaaa \quad &+ bccdaa\end{aligned}}{bd+cd-bcdc+cd+bc} = \frac{- bbccdd}{bd+cd-bc}$$

equal to the sought root a.

75 For if we put b equal to the root a of the proposed equation, then, changing a

into b and expressing the substituted power in terms of the common divisor, we get

$$
\begin{array}{lll}
+\ bdbbbb & +\ bbcbbb & \\
+\ cdbbbb & +\ bccbbb & -\ bbccbb \\
-\ bcdbbb & +\ bddbbb & -\ bbddbb \\
\hline
bd + cd - bc & -\ cddbbb & -\ bcddbb \\
 & -\ bbdbbb & -\ ccddbb \\
 & -\ ccdbbb & +\ bbcdbb \\
 & -\ 2bcdbbb & +\ bccdbb \quad = \quad -\ bbccdd \\
\hline
 & \overline{bd + cd - bc}\ \overline{bd + cd - bc} & \overline{bd + cd - bc}
\end{array}
$$

But this equality is obvious, once terms of opposite signs have been removed.

Thus having put b equal to the root a, it is indeed [shown to be] equal.

Likewise, if we put c equal to the root a, a similar equality follows from the substitution; and thus we must similarly conclude that that too is a root equal to the root a.

Thus b and c are roots equal to the sought root a, as was stated in the proposition.

PROPOSITION 35

b and c are the roots of the equation
$$
\begin{array}{l}
aaaa - bbba \\
\quad\ - bbca \\
\quad\ - bcca \\
\quad\ - ccca = - bbbc \\
\quad\qquad\qquad\ - bbcc \\
\quad\qquad\qquad\ - bccc
\end{array}
$$
equal to the sought root a.

For if we put $b = a$ we get
$$
\begin{array}{l}
bbbb - bbbb \\
\quad\ - bbbc \\
\quad\ - bbcc \\
\quad\ - bccc = - bbbc \\
\quad\qquad\qquad - bbcc \\
\quad\qquad\qquad - bccc
\end{array}
$$

Or if we put $c = a$ we get
$$
\begin{array}{l}
cccc - bbbc \\
\quad\ - bbcc \\
\quad\ - bccc \\
\quad\ - cccc = - bbbc \\
\quad\qquad\qquad - bbcc \\
\quad\qquad\qquad - bccc
\end{array}
$$

These equalities are obvious.

Thus the root of the proposed equation is $a = b$ or c, as was stated in the proposition.

PROPOSITION 36

b and c are the roots of the equation
$$
\frac{aaaa - bbbaaa}{\quad\ - bbcaaa} \\
$$

$$
\frac{aaaa - bbbaaa - bbcaaa - bccaaa - cccaaa}{bb + bc + cc} = \frac{- bbbccc}{bb + bc + cc}
$$

equal to the sought root a.

For if we put $b = a$, we get

$$
\begin{array}{ll}
+\, bbbbbb & -\, bbbbbb \\
+\, bbbbbc & -\, bbbbbc \\
+\, bbbbcc & -\, bbbbcc \\
\hline
bb + bc + cc - bbbccc & = -\, bbbccc \\
\hline
bb + bc + cc & bb + bc + cc
\end{array}
$$

Or if we put $c = a$, we get

$$
\begin{array}{ll}
+\, bbcccc & -\, bbbccc \\
+\, bccccc & -\, bbccccc \\
+\, cccccc & -\, bccccc \\
\hline
bb + bc + cc - cccccc & = -bbbccc \\
\hline
bb + bc + cc & bb + bc + cc
\end{array}
$$

These equalities are obvious.

Thus the root of the proposed equation is $a = b$ or c, as was stated in the proposition.

PROPOSITION 37

b and c are the roots of the equation $\begin{aligned} aaaa - bbaa \\ -ccaa \end{aligned} = -bbcc$ equal to the sought root a.

For if we put $b = a$, we get $\begin{aligned} bbbb - bbbb \\ - bbcc \end{aligned} = -bbcc$

Or if we put $c = a$, we get $\begin{aligned} cccc - bbcc \\ - cccc \end{aligned} = -bbcc$

These equalities are obvious.

Thus the root of the proposed equation is $a = b$ or c, as was stated in the proposition.

Reciprocal Equations

PROPOSITION 38

b is the root of the equation $aaaa - baaa + cdfa = +bcdf$, equal to the sought root a.

77 For if we put b equal to the root a of the equation $aaaa - baaa + cdfa = +bcdf$, then, changing a into b we get $bbbb - bbbb + cdfb = +cdfb$.

But this equality is obvious in and of itself.

Thus, having put b equal to the root a, it is indeed [shown to be] equal, as was stated in the proposition.

PROPOSITION 39

c is the root of the equation $aaaa + baaa - ccca = +bccc$, equal to the sought root a.

For if we put c equal to the root a of the equation $aaaa + baaa - ccca = +bccc$, then, changing a into c we get $cccc + bccc - cccc = +bccc$.

But this equality is obvious in and of itself.

Thus, having put c equal to the root a, it is indeed [shown to be] equal, as was stated in the proposition.

PROPOSITION 40

b and c are the roots of the equation $aaaa - baaa - ccca = -bccc$ equal to the sought root a.

For if we put b equal to the root a of the equation $aaaa - baaa - ccca = -bccc$, then, changing a into b we get $bbbb - bbbb - cccb = -bccc$

But the truth of this equality is obvious in and of itself.

Thus, having put b equal to the root a, it is indeed [shown to be] equal.

Likewise, if we put c equal to the root a, then, changing a into c we get $cccc - bccc - cccc = -bccc$

But the truth of this equality is obvious.

Thus, if we put c equal to the root a, it is indeed [shown to be] equal.

Thus b and c are roots equal to the sought root a, as was stated in the proposition.

Section Five[1]

The fifth section, in which the number of roots of ordinary equations is determined by means of the equipollence[2] of canonical equations

DEFINITION

Two equations will henceforth be called *equipollent* if they are of the same degree and similarly affected, and if the coefficient or coefficients (if there are several) and the given homogeneous terms of one equation are capable of the simple relation of inequality (that is, greater than or less than) with the coefficient or coefficients and the given homogeneous term of the other. And the term should be further interpreted to mean that the equations possess an equal number of roots.*

This is why the term 'canonical' was applied to equations generated from binomial roots and their reductions – the equations, that is, which were treated in the previous three sections: because they are compared to ordinary equations, and if the conditions of equipollence (described above) hold between them, then they are 'canons', or sure and regular models for discovering and determining the number of roots in the ordinary equations. In order, then, to make the coefficients and given homogeneous terms of ordinary and canonical equations agree with each other, the coefficients and homogeneous terms of the ordinary equations should be distributed in a way similar to the formal distribution of canonical equations, bringing similar parts together on each side and keeping to the law of homogeneity in judging the relationships of these terms; that is, by reduction by an introduced homogeneous term. This is because the coefficients and given homogeneous terms are necessarily heterogeneous, and no judgement can be made of the properties concerning the relationship to each other of heterogeneous terms.

---

*i.e., the Latin word *aequipollentes* should be expanded to the phrase *aequali radicum numero pollentes*.

Lemma 1[3]

If a quantity be divided into two unequal parts, the square of half the total is greater than the product of the two unequal parts.

If p and q are two unequal parts of the magnitude, then it is true that

$$\left.\begin{array}{c}\frac{p+q}{2}\\[2pt]\frac{p+q}{2}\end{array}\right| > pq$$

For, from three (quantities) in continued proportion pp, pq and qq, of which pp is the greatest and qq the least, it is true that $pp - pq > pq - qq$.

Therefore $pp + qq > 2pq$.

And, adding $2pq$ to both sides, it is true that $pp + 2pq + qq > 4pq$.

But $pp + 2pq + qq = 1\left.\begin{array}{c}p+q\\p+q\end{array}\right|$

Therefore $\left.\begin{array}{c}p+q\\p+q\end{array}\right| > 4pq$.

Therefore $\dfrac{\left.\begin{array}{c}p+q\\p+q\end{array}\right|}{4} > pq$.

79 Therefore

$$\left.\begin{array}{c}\frac{p+q}{2}\\[2pt]\frac{p+q}{2}\end{array}\right| > pq$$

which was to be proved.

Lemma 2

If three quantities are in continued proportion, the sum of the extremes is greater than twice the middle.

Suppose b, c and d are in continued proportion; then it is true that $b + d > 2c$.

For, if b is the greatest, then $b - c > c - d$.

Therefore $b + d > 2c$.

Or if d is the greatest, then $d - c > c - b$.

Therefore $d + b > 2c$.

Consequently the sum of the extremes is greater than twice the middle, as was stated in the lemma.

Lemma 3[4]

If four quantities are in continued proportion, the sum of the extremes is greater than the sum of the middle terms.

If b, c, d and f are in continued proportion, it is true that $b + f > c + d$.

For, if b is the greatest, then $b - c > d - f$.

Therefore $b + f > c + d$.

Or if f is the greatest, then $f - d > c - b$.

Therefore $f + b > c + d$.

Consequently the sum of the extremes is greater than the sum of the middle terms, as was stated in the lemma.

PROPOSITION 1[5]

The ordinary equation $aaa - 3bba = +2ccc$ in which $c > b$, is explicable in terms of a single root.

For, the proposed ordinary equation is of the same degree and similarly affected to the canonical equation $aaa - 3rqa = +rrr + qqq$.

And (by the following Lemma 4) in the canonical equation it is true that   **80**

$$\left.\begin{array}{c} rq \\ rq \\ \underline{rq} \end{array}\right| < \left.\frac{\begin{array}{c} rrr + qqq \\ rrr + qqq \end{array}}{4}\right|$$

And in the proposed equation, in which it is supposed that $b < c$, it is true that $bbbbbb < cccccc$.

Therefore the coefficient and the given homogene of the proposed equation conform to the coefficient and given homogene of the canonical equation, in the relationship of excess and defect.

Accordingly (by the Definition) the proposed equation and the canonical equation are equipollent (that is, are provided with an equal number of roots).

But (by Prop. 14 of Section 4) the canonical equation is explicable in terms of the single root $q + r$.

And so the proposed ordinary equation is explicable in terms of a single root, as was stated in the proposition.

LEMMA 4[6]

It is true that $\dfrac{\left.\begin{array}{c} rrr + qqq \\ rrr + qqq \end{array}\right|}{4} > \left.\begin{array}{c} rq \\ rq \\ rq \end{array}\right|$

For, $rrrrrr, rrrqqq$ and $qqqqqq$ are in continued proportion.

Therefore (by Lemma 2) $rrrrrr + qqqqqq > 2rrrqqq$.

And, by adding $2rrrqqq$ to both sides, we get $rrrrrr + 2rrrqqq + qqqqqq > 4rrrqqq$.

Moreover it is true that $rrrrrr + 2rrrqqq + qqqqqq = \left.\begin{array}{c} rrr + qqq \\ rrr + qqq \end{array}\right|$

And $4rrrrqqq = 4.\ \begin{array}{|l} rq \\ rq \\ rq \end{array}$

Therefore $\begin{array}{|l} rrr + qqq \\ rrr + qqq \end{array} > 4.\ \begin{array}{|l} rq \\ rq \\ rq \end{array}$

Therefore $\dfrac{\begin{array}{|l} rrr + qqq \\ rrr + qqq \end{array}}{4} > \begin{array}{|l} rq \\ rq \\ rq \end{array}$ which was to be proved.

PROPOSITION 2[7]

The ordinary equation $aaa - 3bba = +2ccc$, in which $c < b$ is explicable in terms of a single root.

81 For, the proposed ordinary equation is of the same degree and similarly affected as the canonical equation
$$aaa \begin{array}{l} -qqa \\ -qra \\ -rra \end{array} = \begin{array}{l} +qqr \\ +qrr \end{array}$$

And (by the following Lemma 5) in the canonical equation it is true that
$$\dfrac{\begin{array}{|l} qqr + qrr \\ qqr + qrr \end{array}}{4} < \dfrac{\begin{array}{|l} qq + qr + rr \\ qq + qr + rr \\ qq + qr + rr \end{array}}{27}$$

And in the proposed equation, in which it is supposed that $c < b$, it is true that $cccccc < bbbbbb$.

Therefore the coefficient and given homogene of the proposed equation conform to the coefficient and given homogene of the canonical equation in the relationship of inequality.

Accordingly (by the Definition) the proposed equation and the canonical equation are equipollent (that is, are provided with an equal number of roots).

But (by Proposition 7 of Section 4) the canonical equation is explicable in terms of a single root $q + r$.

Consequently the proposed ordinary equation is explicable in terms of a single root, as was stated in the proposition.

LEMMA 5[8]

It is true that $\dfrac{\begin{array}{|l} qqr + qrr \\ qqr + qrr \end{array}}{4} < \dfrac{\begin{array}{|l} qq + qr + rr \\ qq + qr + rr \\ qq + qr + rr \end{array}}{27}$

For, by Lemma 3, it is true that $3\,qqqqqr + 3\,qqrrrr < 3\,qqqqqq + 3\,rrrrrr$;
and, by Lemma 2, $12\,qqqrrr + 12\,qqqrrr < 12\,qqqqqr + 12\,qrrrrr$;
and, by Lemma 2, $qqqrrr + qqqqrrr < qqqqqq + rrrrrr$.

Therefore $\begin{array}{l} +\ 3\,qqqqrr \\ +26\,qqqqrrr \\ +\ 3\,qqrrrr \end{array} < \begin{array}{l} +\ 4\,qqqqqq \\ +12\,qqqqqr \\ +\ 12\,qrrrrr \\ +\ 4\,rrrrrr \end{array}$

Therefore, adding to both sides $24\,qqqqrr + 28\,qqqrrr + 24\,qqrrrr$, it is true that

$$
\begin{array}{c}
+27\,qqqqrr \\
+54\,qqqrrr \\
+27\,qqrrrr
\end{array}
<
\begin{array}{l}
+\ 4\,qqqqqq \\
+12\,qqqqqr \\
+\ 12\,qrrrrr \\
+\ \ 4\,rrrrrr \\
+24\,qqqqrr \\
+28\,qqqrrr \\
+\ 24\,qqrrrr
\end{array}
$$

Then, dividing the terms on both sides throughout by 4, and again throughout by 27,

it is true that
$$
\frac{
\begin{array}{l}
+1\,qqqqqrr \\
+2\,qqqqrrr \\
+\ 1\,qqrrrrr
\end{array}
}{4}
<
\frac{
\begin{array}{l}
+1\,qqqqqq \\
+3\,qqqqqr \\
+\ 3\,qrrrrr \\
+\ 1\,rrrrrr \\
+6\,qqqqrr \\
+7\,qqqrrr \\
+\ 6\,qqrrrr
\end{array}
}{27}
$$

But
$$
\frac{
\begin{array}{l}
+1\,qqqqrr \\
+2\,qqqrrr \\
+\ 1\,qqrrrr
\end{array}
}{4}
=
\frac{
\begin{array}{l}
+qqr + qrr \\
+qqr + qrr
\end{array}\Big|}{4}
$$

and
$$
\frac{
\begin{array}{l}
+1\,qqqqqq \\
+3\,qqqqqr \\
+\ 3\,qrrrrr \\
+\ 1\,rrrrrr \\
+6\,qqqqrr \\
+7\,qqqrrr \\
+\ 6\,qqrrrr
\end{array}
}{27}
=
\frac{
\begin{array}{l}
+qq + qr + rr \\
+qq + qr + rr \\
+qq + qr + rr
\end{array}\Big|}{27}
$$

Therefore
$$
\frac{
\begin{array}{l}
qqr + qrr \\
qqr + qrr
\end{array}\Big|}{4}
<
\frac{
\begin{array}{l}
qq + qr + rr \\
qq + qr + rr \\
qq + qr + rr
\end{array}\Big|}{27}
$$
which was to be proved.

PROPOSITION 3

The ordinary equation $aaa - 3bba = +2ccc$, in which $c = b$, is explicable in terms of a single root.

For, the proposed ordinary equation is of the same degree and similarly affected as the canonical equation $aaa - 3qqa = +2qqq$.

And in the canonical equation, it is true that $q = q$.

Moreover, in the proposed equation, it is supposed that $b = c$.

Therefore the coefficient and given homogene of the proposed equation conform to the coefficient and given homogene of the canonical equation in their condition of equality: for they are equal on both sides (making the comparison in correct accordance with the law of homogeneity).

Consequently the proposed equation and the canonical are equipollent, that is, are provided with an equal number of roots.

But (by Proposition 17 of Section 4) the canonical equation is explicable in terms of a single root $2q$.

Consequently the proposed ordinary equation is to be explicated in terms of a single root, as was stated in the proposition.

PROPOSITION 4

83 The ordinary equation $aaa - 3bba = -2ccc$, in which $b > c$, is explicable in terms of two roots.

For, the proposed ordinary equation is of the same degree and similarly affected as the canonical equation

$$
\begin{array}{l}
aaa -qqa \\
\quad\ -qra \\
\quad\ -rra = -qqr \\
\quad\qquad\quad -qrr
\end{array}
$$

And (by Lemma 5 to Proposition 2) in the canonical equation, it is true that

$$
\left.\begin{array}{c} qq + qr + rr \\ qq + qr + rr \\ qq + qr + rr \end{array}\right| \over 27 \quad > \quad \left.\begin{array}{c} +qqr \\ +qrr \end{array}\right| \over 4
$$

And in the proposed equation, in which it is supposed that $b > c$, it is true that $bbbbbbb > ccccc$.

Therefore the coefficient and given homogene of the proposed equation conform to the coefficient and given homogene of the canonical equation, in the relationship of excess and defect.

Accordingly (by the Definition) the canonical equation and the proposed equation are equipollent (that is, endowed with an equal number of roots).

But (by Proposition 6 of Section 4) the canonical equation is explicable in terms of the two roots q and r.

And so the proposed equation is explicable in terms of two roots, as was stated in the proposition.

PROPOSITION 5[9]

The ordinary equation $aaa - 3baa + cca = +ddd$, in which $b > c$ and $b > d$, is explicable in terms of three roots.

For, the proposed ordinary equation is of the same degree and similarly affected as the canonical equation

$$
\begin{array}{l}
aaa +paa +pqa \\
\quad\ -qaa +pra \\
\quad\ -raa +qra = +pqr
\end{array}
$$

And in the canonical equation (by the following Lemma 6) it is true that

$$
\left.\begin{array}{c} \frac{p+q+r}{3} \\[4pt] \frac{p+q+r}{3} \end{array}\right| \quad > \quad \frac{pq+pr+qr}{3}
$$

And (by the following Lemma 7)   **84**

$$\left.\begin{array}{c} \frac{p+q+r}{3} \\ \frac{p+q+r}{3} \\ \frac{p+q+r}{3} \end{array}\right| > pqr$$

Moreover, in the proposed equation (in which it is supposed that $b > c$, and $b > d$), it is true that $bb > cc$ and $bbb > ddd$.

Therefore the coefficient and the given homogene of the proposed equation conform to the coefficient and given homogene of the canonical equation, in the relationship of excess and defect.

Accordingly (by the Definition) the proposed equation and the canonical are equipollent (that is, endowed with an equal number of roots).

But (by Proposition 5 of Section 4) the canonical equation is explicable in terms of the three roots p, q and r.

And so the proposed ordinary equation is explicable in terms of three roots, as was stated in the proposition.

LEMMA 6[10]

If a quantity be divided into three unequal parts, the square on one-third of the whole is greater than one-third of the [sum of all] the products of the unequal parts taken in pairs.

If the three unequal parts of the quantity are p, q and r, then it is true that

$$\left.\begin{array}{c} \frac{p+q+r}{3} \\ \frac{p+q+r}{3} \end{array}\right| > \frac{pq+pr+qr}{3}$$

For (by Lemma 2), $pp + qq > 2pq$;
and $qq + rr > 2qr$;
and $pp + rr > 2pr$.

Therefore $2pp + 2qq + 2rr > 2pq + 2qr + 2pr$.

Therefore $pp + qq + rr > pq + qr + pr$.

And, adding $2pq + 2qr + 2pr$ to both sides, it will be true that
$$\begin{array}{r} pp + qq + rr \\ +2pq + 2qr \\ +2pr \end{array} > 3pq + 3qr + 3pr$$

But $\left.\begin{array}{c} pp + qq + rr \\ +2pq + 2qr \\ +2pr \end{array}\right. = \left.\begin{array}{c} p+q+r \\ p+q+r \end{array}\right|$

Therefore $\left.\begin{array}{c} p+q+r \\ p+q+r \end{array}\right| > 3pq + 3qr + 3pr.$

That is, $\dfrac{\begin{array}{c}p+q+r\\p+q+r\end{array}}{3} > pq + qr + pr.$

85 Therefore,

$$\dfrac{\begin{array}{c}\frac{p+q+r}{3}\\\frac{p+q+r}{3}\end{array}}{} > \dfrac{pq+pr+qr}{3}$$

which was to be proved.

LEMMA 7[11]

If a quantity be divided into three unequal parts, the cube on one-third of the whole is greater than the solid made from the three unequal parts.

If the three unequal parts of the quantity are p, q and r, then it is true that

$$\begin{array}{c}\frac{p+q+r}{3}\\\frac{p+q+r}{3}\\\frac{p+q+r}{3}\end{array} > pqr$$

For (by Lemma 2), $pp + qq > 2pq$;
and $qq + rr > 2qr$;
and $pp + rr > 2pr$.

Therefore $ppr + qqr > 2pqr$;
and $pqq + prr > 2pqr$;
and $ppq + qrr > 2pqr$.

Therefore $\begin{array}{c}+pqq + prr\\+ppq + qrr\\+ppr + qqr\end{array} > 6pqr.$

But (by Lemma 3), $ppp + qqq > ppq + pqq$;
and $qqq + rrr > qqr + qrr$;
and $ppp + rrr > ppr + prr$.

Therefore $ppp + qqq + rrr > 3pqr$.

And $\begin{array}{c}+3pqq + 3prr\\+3ppq + 3qrr\\+3ppr + 3qqr\end{array} > 18pqr.$

Therefore $\begin{array}{c}ppp + qqq + rrr\\+3pqq + 3prr\\+3ppq + 3qrr\\+3ppr + 3qqr\end{array} > 21pqr.$

And, adding $6pqr$ to both sides, it is true that $\begin{array}{c}ppp + qqq + rrr\\+3pqq + 3prr\\+3ppq + 3qrr\\+3ppr + 3qqr\\+6pqr\end{array} > 27pqr.$

86 But $\begin{array}{c}ppp + qqq + rrr\\+3pqq + 3prr + 3ppq\\+3qrr + 3ppr + 3qqr\\+6pqr\end{array} = \begin{array}{c}p + q + r\\p + q + r\\p + q + r\end{array}$

Therefore $\left. \begin{array}{c} p+q+r \\ p+q+r \\ p+q+r \end{array} \right| > 27pqr.$

That is, $\dfrac{\left. \begin{array}{c} p+q+r \\ p+q+r \\ p+q+r \end{array} \right|}{27} > pqr.$

Therefore,

$\left. \begin{array}{c} \frac{p+q+r}{3} \\ \frac{p+q+r}{3} \\ \frac{p+q+r}{3} \end{array} \right| > pqr$

which was to be demonstrated.

PROPOSITION 6[12]

The ordinary equation $aaaa - 4bbba = -3cccc$, in which $b > c$, is explicable in terms of a two roots.

For, the proposed ordinary equation is of the same degree and similarly affected as the canonical equation
$$aaaa - bbba$$
$$-bbca$$
$$-bcca$$
$$-ccca = -bbbc$$
$$\qquad\quad -bbcc$$
$$\qquad\quad -bcc$$

And in the canonical equation the fourth power of $\frac{bbb+bbc+bcc+ccc}{4}$ is greater than the cube of $\frac{bbbc+bbcc+bccc}{3}$.

And in the proposed equation, in which it is supposed that $b > c$, the fourth power of $\frac{bbb}{4}$ is greater than the cube of $\frac{cccc}{3}$.

Therefore the coefficient and the given homogene of the proposed equation conform to the coefficient and given homogene of the canonical equation, in the relationship of excess and defect.

Accordingly (by the Definition) the proposed equation and the canonical are equipollent (that is, are endowed with an equal number of roots).

But (by Proposition 35 of Section 4) the canonical equation is explicable in terms of two roots, b and c

Consequently the proposed equation is explicable in terms of two roots, as was stated in the proposition.

The reduction of ordinary equations by the exclusion of a lower power and a change of the substituted root

A problem on multiplying the roots of equations, useful for preparing those equations for reduction, whose reductions are presented in this Section

To multiply the root of a given equation by any given number whatsoever, while preserving the equality of comparison.[2]

Suppose we have the quadratic equation $aa + ba = ca$

Then $\left.\dfrac{1}{aa}\right| + \left.\dfrac{b}{a}\right| = \left.\dfrac{cc}{1}\right|$

But because the root of the equation is to be doubled, first its three homogeneous terms must be multiplied by three numbers in double proportion (1, 2, 4) applied in order.

Which gives $\begin{array}{c} 1 \\ 1 \\ \hline aa \end{array}\Bigg| + \begin{array}{c} 2 \\ b \\ a \end{array}\Bigg| < \begin{array}{c} 4 \\ cc \\ 1 \end{array}\Bigg|$

(since the equality is destroyed).

Consequently, to restore the equality, the homogeneous terms of the equation must be again multiplied by proportional numbers, but applied in reverse order (that is, 4, 2, 1).

Which will give $\begin{array}{c} 1 \\ 1 \\ aa \\ \hline 4 \end{array}\Bigg| + \begin{array}{c} 2 \\ b \\ a \\ 2 \end{array}\Bigg| = \begin{array}{c} 4 \\ cc \\ 1 \\ 1 \end{array}\Bigg|$

For these are equal: $\left.\dfrac{1}{4}\right| = \left.\dfrac{2}{2}\right| = \left.\dfrac{4}{1}\right|$

Now let $2a = e$.

Then $\left.\dfrac{aa}{4}\right| \& \left.\dfrac{a}{2}\right| \& \left.\dfrac{1}{1}\right| = cc \ \& \ e \ \& \ 1.$

And so, substituting these into their places, we get $\begin{array}{c|c|c} \frac{1}{ee} & \frac{2}{e} & \frac{4}{1} \\ 1 & b & cc \end{array}$

Therefore $ee + 2be = 4cc$.

And so, the root a of the proposed equation is doubled by changing $2a$ into e, while preserving equality; as was required.

88 To multiply the root of a cubic equation.

Let $aaa + baa + cca = ddd$ be a cubic equation of which the root a is to be tripled.

Since $aaa + baa + cca = ddd$, then $\begin{array}{c|c|c|c} \frac{1}{aaa} & \frac{b}{aa} & \frac{cc}{a} & \frac{ddd}{1} \\ 1 & & & \end{array}$

But because the root of the equation is to be tripled, let its four homogeneous terms be multiplied by four numbers in triple proportion $(1, 3, 9, 27)$ applied in order.

Which gives $\begin{array}{c|c|c|c} \frac{1}{aaa} & \frac{3}{aa} & \frac{9}{aa} & \frac{27}{1} \\ 1 & b & cc & ddd \end{array} <$

(since the equality is destroyed).

Consequently, to restore the equality, the homogeneous terms of the equation must be again multiplied by proportional numbers, but applied in reverse order (that is, $27, 9, 3, 1$).

Which will give $\begin{array}{c|c|c|c} \frac{1}{aaa} & \frac{3}{aa} & \frac{9}{a} & \frac{27}{1} \\ \frac{1}{27} & \frac{b}{9} & \frac{cc}{3} & \frac{ddd}{1} \end{array} =$

For these are equal: $\frac{1}{27} = \frac{3}{9} = \frac{9}{3} = \frac{27}{1}$

Now let $3a = e$.

Then $\frac{aaa}{27}$ & $\frac{aa}{9}$ & $\frac{a}{3}$ & $\frac{1}{1}$ = eee & ee & e & 1.

And so, substituting these into their places, we get $\begin{array}{c|c|c|c} \frac{1}{eee} & \frac{3}{ee} & \frac{9}{e} & \frac{27}{1} \\ 1 & b & cc & ddd \end{array} =$

Therefore $eee + 3bee + 9cce = 27ddd$.

And so, the root a of the proposed equation is tripled by changing $3a$ into e; as was required.

Conclusion

The multiplication of a root entails the multiplication of the simple coefficient by the same degree of multiplicity. Thus, in the previous examples, in a quadratic equation whose root a is doubled, the coefficient b is multiplied likewise by 2; and in a cubic equation whose root a is tripled, the co-efficient b is multiplied likewise by ternary numbers. Furthermore, if a problem were conceived and proposed concerning the multiplication of a coefficient, it would be equipollent[3] to this, only with the direction reversed. For the multiplication of a coefficient presupposes the multiplication of the root. This inevitable multiplication of a coefficient is of great use in resolving equations where fractions need to be **89** removed, or in handling the reductions of equations where the indivisibility of the coefficient (whether a number or a symbol) is an obstacle – here one can use it to avoid getting tied up in fractions; so useful is it that, for this reason alone, this seems to be the principal application of this technique.

PROBLEM 1[4]

To reduce the equation $aaa - 2baa = +ccc$ to the equation $eee - 3bbe = +ccc + 2bbb$ by putting $a = c + b$, or to the impossible equation $eee - 3bbe = -ccc - 2bbb$ by putting $a = -e + b$.

Let us first put $e + b = a$.

And let it become $ee + 2be + bb = aa$.

$$\left.\begin{array}{ll} \text{And} & eee + 3bee + 3bbe + bbb = +aaa \\ \text{And also} & -3bee - 6bbe - 3bbb = -3baa \end{array}\right\} = +ccc$$

Removing the terms that cancel and reordering the rest, it becomes:

$$eee - 3bbe = +ccc \\ +2bbb$$

the first sought equation of which the root is $e = a - b$.

Next, let us put $-e + b = a$.

And let it become $ee - 2be + bb = aa$.

$$\left.\begin{array}{ll} \text{And} & -eee + 3bee - 3bbe + bbb = +aaa \\ \text{And also} & -3bee + 6bbe - 3bbb = -3baa \end{array}\right\} = +ccc$$

Removing the terms that cancel and reordering the rest, it becomes:

$$eee - 3bbe = -ccc \\ -2bbb$$

the second sought equation, which is shown to be impossible in the following Lemma.

And in this way is achieved the required reduction of the proposed equation to the sought equations.

LEMMA[5]

The reduced equation $\begin{array}{l} eee - 3bbe = -ccc \\ -2bbb \end{array}$ is impossible.

For, putting $e = b$ it will be true that $\begin{array}{l} bbb - 3bbb = -ccc \\ -2bbb \end{array}$ Therefore $ccc = 0$, which is impossible.

Or, putting $e > b$ (i.e., $e = b + d$) it will be true that $\begin{array}{l} +bbb \\ +3bbd \\ +3bdd - 3bbb \\ +ddd - 3bbd = -ccc \\ - 2bbb \end{array}$ **90**

Removing the terms that cancel, we get $\begin{array}{l} +3bdd \\ +ddd = -ccc \end{array}$

Therefore $\begin{array}{l} +3bdd \\ +ddd \\ +ccc = 0 \end{array}$ which is also impossible.

Or putting $e < b$ (i.e., $e = b - d$) it will be true that $\begin{array}{l} +bbb \\ -3bbd \\ +3bdd - 3bbb \\ -ddd + 3bbd \end{array} = \begin{array}{l} -ccc \\ -2bbb \end{array}$

Removing the terms that cancel, we get $\begin{array}{l} +ddd \\ -3bdd \end{array} = +ccc$

But $+ddd - 3bdd = \begin{array}{|l|} +d - 3b \\ \hline dd \end{array}$

Therefore $d > 3b$; and therefore $d > b$.

But since we have put $e = b - d$, then $b > d$, which is also impossible.

And so the reduced equation is impossible in all cases.

PROBLEM 2[6]

To reduce the equation $aaa + 3baa = +ccc$, to the equation $eee - 3bbe = \begin{array}{l} +ccc \\ -2bbb \end{array}$ by putting $a = e - b$.

Let us put $e - b = a$.

Which gives $eee \quad - 3bee + 3bbe - \quad bbb = \quad +aaa$
And $\qquad\qquad ee - \quad 2be + \quad bb = \quad aa \left.\begin{array}{r}\\\\\end{array}\right\} = +ccc$
And consequently $+ 3bee - 6bbe + 3bbb = +3baa$

Removing the terms that cancel and reordering the rest, it becomes:

$$eee - 3bbe = +ccc$$
$$-2bbb$$

the sought equation, of which the root is $e = a + b$.

And in this way is achieved the reduction of the proposed equation to the sought equation.

PROBLEM 3[7]

91 To reduce the equation $aaa - 3baa = -ccc$ to the equation
$eee - 3bbe = \begin{array}{l} +ccc \\ -2bbb \end{array}$ by putting $a = b - e$; or to the equation
$eee - 3bbe = \begin{array}{l} -ccc \\ +2bbb \end{array}$ by putting $a = e + b$.

First let us put $b - e = a$.

Which gives $\quad - eee + 3bee - 3bbe + \quad bbb = \quad +aaa$
And $\qquad\qquad ee - \quad 2be + \quad bb = \quad aa \left.\begin{array}{r}\\\\\end{array}\right\} = -ccc$
And consequently $\quad - 3bee + 6bbe - 3bbb = -3baa$

Removing the terms that cancel and reordering the rest, it becomes:

$$eee - 3bbe = +ccc$$
$$-2bbb$$

the first sought equation, of which the root is $e = b - a$.

Next, let us put $e + b = a$.

$$\left.\begin{array}{l}\text{Which gives } eee \quad + 3bee + 3bbe + \ bbb = \ +aaa \\ \text{And} \qquad\qquad\quad + \ ee + \ 2be + \ bb = \qquad aa \\ \text{And consequently } - 3bee - 6bbe - 3bbb = -3baa\end{array}\right\} = -ccc$$

Removing the terms that cancel and reordering the rest, it becomes:

$$\begin{array}{l} eee - 3bbe = -ccc \\ \qquad\quad +2bbb \end{array}$$

the second sought equation, of which the root is $e = a - b$.

And in this way is achieved the required reduction to both the sought equations.

PROBLEM 4[8]

To reduce the equation $aaa + 3baa + dda = +ccc$ to the equation

$$\begin{array}{l} eee - 3bbe \\ \quad + dde \ = +ccc \\ \qquad\qquad -2bbb \\ \qquad\quad +bdd \end{array} \quad \text{by putting } a = e - b.$$

First let us put $e - b = a$.

$$\left.\begin{array}{l}\text{Which gives } eee \quad - \quad 3bee + 3bbe - \ bbb = \ +aaa \\ \text{And} \qquad\qquad\qquad\quad ee - \quad 2be + \ bb = \qquad aa \\ \text{And consequently} \quad +3bee - \quad 6bbe + 3bbb = +3baa \\ \text{And} \qquad\qquad\qquad\quad +dde - \ ddb = \quad dda\end{array}\right\} = +ccc$$

Hence it becomes: **92**

$$\begin{array}{l} eee - 3bbe \\ \quad + dde \ = +ccc \\ \qquad\qquad -2bbb \\ \qquad\quad +bdd \end{array}$$

the sought equation, of which the root is $e = a + b$.

And in this way is achieved the required reduction.

PROBLEM 5[9]

To reduce the equation $aaa - 3baa + dda = -ccc$ to the equation

$$\begin{array}{l} eee - 3bbe \\ \quad + dde \ = +ccc \\ \qquad\qquad -2bbb \\ \qquad\quad +bdd \end{array} \quad \text{by putting } a = b - e;$$

or to the equation
$$\begin{array}{l} eee - 3bbe \\ \quad + dde \ = -ccc \\ \qquad\qquad +2bbb \\ \qquad\quad -bdd \end{array} \quad \text{by putting } a = e + b.$$

First let us put $-e + b = a$.

Which gives $ee - 2be + bb = aa$.

Which gives $\quad -eee + 3bee - 3bbe + \ bbb = \ +aaa$

And $\qquad\qquad -3bee + 6bbe - 3bbb = -3baa \left.\vphantom{\begin{matrix}a\\b\\c\end{matrix}}\right\} = -ccc$

And $\qquad\qquad\qquad -dde + bdd = +dda$

Removing the terms that cancel and reordering and transposing the rest, it becomes:

$$eee - 3bbe$$
$$+dde = +ccc$$
$$-2bbb$$
$$+bdd$$

the first sought equation, of which the root is $e = b - a$.

Next, let us put $e + b = a$.

Which gives $ee + 2be + bb = aa$.

Which gives $eee + 3bee + 3bbe + \ bbb = \ +aaa$

And $\qquad\qquad -3bee - 6bbe - 3bbb = -3baa \left.\vphantom{\begin{matrix}a\\b\\c\end{matrix}}\right\} = -ccc$

And $\qquad\qquad\qquad + dde + bdd = +dda$

Removing the terms that cancel and reordering the rest, it becomes:

$$eee - 3bbe$$
$$+dde = -ccc$$
$$+2bbb$$
$$-bdd$$

the second sought equation, of which the root is $e = a - b$.

93 And in this way is achieved the required reduction to both the sought equations.

PROBLEM 6[10]

To reduce the equation $aaa + 3baa - dda = +ccc$ to the equation

$$eee - 3bbe$$
$$-dde = +ccc$$
$$-2bbb \quad \text{by putting } a = e - b.$$
$$-bdd$$

Let us put $e - b = a$.

Which gives $ee - 2be + bb = aa$.

Which gives $eee - 3bee + 3bbe - \ bbb = \ +aaa$

And $\qquad\qquad +3bee - 6bbe + 3bbb = +3baa \left.\vphantom{\begin{matrix}a\\b\\c\end{matrix}}\right\} = +ccc$

And $\qquad\qquad\qquad - dde + bdd = -dda$

Removing the terms that cancel and reordering and transposing the rest, it becomes: $\ eee - 3bbe$

$$-dde = +ccc$$
$$-2bbb$$
$$-bdd$$

the reduced equation, of which the root is $e = a + b$.

And in this way is achieved the reduction of the proposed equation to the sought equation.

PROBLEM 7[11]

To reduce the equation $aaa - 3baa - daa = -ccc$ to the equation

$$eee - 3bbe$$
$$- dde = +ccc$$
$$-2bbb \quad \text{by putting } a = b - e;$$
$$-bdd$$

or to the equation

$$eee - 3bbe$$
$$- dde = -ccc$$
$$+2bbb \quad \text{by putting } a = b + e.$$
$$+bdd$$

Let us put $e + b = a$.

Which gives $ee - 2be + bb = aa$.

94

Which gives $\left. \begin{array}{l} - eee + 3bee - 3bbe + bbb = +aaa \\ - 3bee + 6bbe - 3bbb = -3baa \\ + dde - bdd = -dda \end{array} \right\} = -ccc$

And

And

Removing the terms that cancel and reordering and transposing the rest, it becomes: $eee - 3bee$

$$- dde = +ccc$$
$$-2bbb$$
$$-bda$$

the first reduced equation, of which the root is $e = +b - a$.

Next, let us put $e + b = a$.

Which gives $ee + 2be + bb = aa$.

Which gives $\left. \begin{array}{l} eee + 3bee + 3bbe + bbb = +aaa \\ - 3bee - 6bbe - 3bbb = -3baa \\ + dde - bdd = -dda \end{array} \right\} = -ccc$

And

And

Removing the terms that cancel and reordering the rest, it becomes:

$$eee - 3bbe$$
$$- dde = - ccc$$
$$+ 2bbb$$
$$+ bdd,$$

the second reduced equation of which the root is $e = a - b$

And in this way is achieved the reduction of the proposed equation to the sought equations.

PROBLEM 8[12]

To reduce the equation $aaa - 3baa - dda = +ccc$ to the equation
$$eee - 3bbe$$
$$- dde = + ccc$$
$$+ 2bbb \quad \text{by putting } a = e + b, \text{ or to the equation}$$
$$+ bdd,$$

$$eee - 3bbe$$
$$- dde = + ccc$$
$$ - 2bbb$$ by putting $a = b - e$.
$$ - bdd$$

Let us first put $e + b = a$.

Which gives $ee + 2be + bb = aa$.

Which gives $eee + 3bee + 3bbe + bbb = +aaa$ ⎫
And $ - 3bee - 6bbe - 3bbb = -3baa$ ⎬ $= +ccc$
And $ - dde - bdd = -dda$ ⎭

95

Removing the terms that cancel and reordering the rest, it becomes:

$$eee - 3bbe$$
$$- dde = + ccc$$
$$ + 2bbb$$
$$ + bdd,$$

the first reduced equation of which the root is $e = a - b$.

Next, let us put $-e + b = a$.

Which gives $ee - 2be + bb = aa$.

From which we next get $-eee + 3bee - 3bbe + bbb = +aaa$ ⎫
And $ - 3bee + 6bbe - 3bbb = -3baa$ ⎬ $= +ccc$
And $ + dde - bdd = dda$ ⎭

Removing the terms that cancel and reordering and transposing the rest, it becomes: $eee - 3bbe$
$$- dde = - ccc$$
$$ - 2bbb$$
$$ - bdd,$$

the second sought equation, of which the root is $e = b - a$

And in this way is achieved the reduction of the proposed equation to the sought equations.

PROBLEM 9[13]

To reduce the equation $aaa + 3baa - dda = -ccc$ to the equation
$$eee - 3bbe$$
$$- dde = + ccc$$
$$ + 2bbb$$ by putting $a = -e - b$, or to the equation
$$ + bdd,$$
$$eee - 3bbe$$
$$- dde = - ccc$$
$$ - 2bbb$$ by putting $a = e - b$.
$$ - bdd$$

Let us first put $+e - b = a$.

Which gives $ee + 2be + bb = aa$.

Which gives $-eee-3bee-3bbe-\ bbb=\ +aaa$
And $\qquad\qquad +3bee+6bbe+3bbb=+3baa \Big\} = -ccc$
And $\qquad\qquad\qquad +\ dde+\ bdd=\ -dda$

Removing the terms that cancel and reordering and transposing the rest, it be-

comes: $\quad \begin{aligned}eee-3bbe\\-dde=\begin{matrix}+ccc\\+2bbb\\+bdd,\end{matrix}\end{aligned}$ the first reduced equation of which the root is

$e=-a-b.$

Next, let us put $e-b=a.$ **96**

Which gives $ee-2be+bb=aa.$

Which gives $eee-3bee+3bbe-\ bbb=\ +aaa$
And $\qquad\qquad +3bee-6bbe+3bbb=+3baa \Big\} = -ccc$
And $\qquad\qquad\qquad -\ dde+\ bdd=\ -dda$

Removing the terms that cancel and reordering the rest, it becomes:

$\quad \begin{aligned}eee-3bbe\\-dde=\begin{matrix}-ccc\\-2bbb\\-bdd,\end{matrix}\end{aligned}$ the second sought equation of which the root is

$e=a+b$

And in this way is achieved the reduction of the proposed equation to the sought equations.

PROBLEM 10[14]

To reduce the equation $aaa-3baa+dda=+ccc$ to the equation

$\quad \begin{aligned}eee-3bbe\\+dde=\begin{matrix}+ccc\\+2bbb\\-bdd\end{matrix}\end{aligned}$ by putting $a=e+b$, or to the equation

$\quad \begin{aligned}eee-3bbe\\+dde=\begin{matrix}-ccc\\-2bbb\\+bdd\end{matrix}\end{aligned}$ by putting $a=-e+b.$

Let us first put $e+b=a.$

Which gives $ee+2be+bb=aa.$

Which gives $eee+3bee+3bbe+\ bbb=\ +aaa$
And $\qquad\qquad -3bee-6bbe-3bbb=-3baa \Big\} = +ccc$
And $\qquad\qquad\qquad +\ dde+\ bdd=\ +dda$

Removing the terms that cancel and reordering the rest, it becomes:

$\quad \begin{aligned}eee-3bbe\\+dde=\begin{matrix}+ccc\\+2bbb\\-ddb,\end{matrix}\end{aligned}$

the reduced equation of which the root is $e=a-b$

Next, let us put $-e+b=a.$
Which gives $ee-2be+bb=aa.$

97 Which gives $eee + 3bee - 3bbe + bbb = +aaa$
And $- 3bee + 6bbe - 3bbb = -3baa$ $\Big\} = +ccc$
And $- dde + bdd = +dda$

Removing the terms that cancel and reordering and transposing the rest, it becomes:

$$eee - 3bbe$$
$$+ dde = -ccc$$
$$- 2bbb$$
$$+ bdd,$$

the second reduced equation of which the root is $a = b - e$.

And in this way is achieved the reduction of the proposed equation to the sought equations.

PROBLEM 11[15]

To reduce the equation $aaa + baa + daa = -ccc$ to the equation
$$eee - 3bee$$
$$+ ddd = +ccc$$
$$+ 2bbb \quad \text{by putting } a = -e - b, \text{ or to the equation}$$
$$- bdd,$$
$$eee - 3bbe$$
$$+ dde = -ccc$$
$$- 2bbb \quad \text{by putting } a = e - b.$$
$$+ bdd$$

Let us first put $-e - b = a$.
Which gives $ee + 2be + bb = aa$.

Which gives $- eee - 3bee - 3bbe - bbb = +aaa$
And $+ 3bee + 6bbe + 3bbb = +3baa$ $\Big\} = -ccc$
And $- dde - bdd = +dda$

Removing the terms that cancel and reordering and transposing the rest, it becomes:
$$eee - 3bbe$$
$$+ dde = +ccc$$
$$+ 2bbb \quad \text{the first reduced equation of which the root is}$$
$$- bdd,$$
$e = -a - b$.

Next, let us put $e - b = a$.
Which gives $ee - 2be + bb = aa$.

Which gives $eee - bee + 3bbe - bbb = +aaa$
And $+ 3bee - 6bbe + 3bbb = +3baa$ $\Big\} = -ccc$
And $+ dde - bdd = +dda$

98 Removing the terms that cancel and reordering the rest, it becomes
$$aaa - 3bba$$
$$+ dda = -ccc$$
$$- 2bbb$$
$$+ bdd,$$

the second reduced equation of which the root is $e = a + b$.

And in this way is achieved the reduction of the proposed equation to the sought equations.

PROBLEM 12[16]

To reduce the equation $aaa + 3bba = +2ccc$ to the simple equation $eee = +ccc + \sqrt{ccccc} + bbbbbb$ by putting $a = \frac{ee-bb}{e}$.

Let e, b and $\frac{bb}{e}$ be in continued proportion.

And let the substituted root a of the proposed equation be equal to the difference of the extremes, namely $a = e - \frac{bb}{e}$, i.e., $a = \frac{ee-bb}{e}$

$$\text{Which gives} \quad \left. \begin{array}{c} \dfrac{\begin{array}{c|c|c|c} +ee & -3bb & +3bb & -bb \\ ee & ee & bb & bb \\ ee & ee & ee & bb \end{array}}{eee} = +aaa \\[2em] \text{And} \quad \dfrac{\begin{array}{c|c} +3bb & -3bb \\ ee & bb \\ ee & ee \end{array}}{eee} = +3bba \end{array} \right\} = 2ccc$$

Removing the terms that cancel and reordering the rest, it becomes $+eeeee - bbbbbb = +2ccceee$

Therefore $+eeeee - 2ccceee = +bbbbbb$.

Therefore $+eeeee - 2ccceee + cccccc = +cccccc + bbbbbb$.

Therefore $eee = ccc + \sqrt{ccccc} + bbbbbb$, the equation which was required.

And in this way is achieved the required reduction of the proposed equation to that stated above.

COROLLARY

The resolution of the proposed equation $aaa + 3bba = +2ccc$ can be deduced easily from this very reduction.

For $\overbrace{ccc + \sqrt{ccccc} + bbbbbb}^{I} \quad \overbrace{bbb}^{II} \quad \overbrace{-ccc + \sqrt{ccccc} + bbbbbb}^{III}$ are in **99** continued proportion.

Therefore   so   too   are $\overbrace{\sqrt[3]{ccc + \sqrt{ccccc} + bbbbbb}}^{I} \qquad \overset{II}{b}$

$\overbrace{\sqrt[3]{-ccc + \sqrt{cccccc} + bbbbbb}}^{III}$

But $\sqrt[3]{ccc + \sqrt{cccccc} + bbbbbb} = e$

Therefore $\sqrt[3]{-ccc + \sqrt{cccccc} + bbbbbb} = \frac{bb}{e}$

Therefore $\sqrt[3]{ccc + \sqrt{cccccc} + bbbbbb} - \sqrt[3]{-ccc + \sqrt{cccccc} + bbbbbb} = a$

And so the explicatory root of the proposed equation (which is what had to be expressed) is this binomial radical, itself involving binomial radicals.

Examples of solution in numbers[17]

$$20 = 6a + aaa \quad a = \sqrt[3]{\sqrt{108 + 10}} - \sqrt[3]{\sqrt{108 - 10}} = 2^1$$
$$26 = 9a + aaa \quad a = \sqrt[3]{\sqrt{196 + 13}} - \sqrt[3]{\sqrt{196 - 13}} = 2$$
$$7 = 6a + aaa \quad a = \sqrt[3]{\sqrt{\tfrac{81}{4}} + \tfrac{7}{2}} - \sqrt[3]{\sqrt{\tfrac{81}{4}} - \tfrac{7}{2}} = 1$$

PROBLEM 13[18]

Given the equation $aaa - 3bba = +2ccc$ and the substitution $a = \frac{ee+bb}{e}$ if c, to reduce the given equation (1) to the simple equation $eee = ccc + ddd$, if c is greater than b; (2) to another simple equation $eee = ccc$ if $c = b$; or (3) to the impossible equation $eee = ccc + \sqrt{-dddddd}$ if c is less than b.

Let e, b, $\frac{bb}{e}$ be in continued proportion.

And let the root a of the proposed equation be equal to the sum of the extremes, that is, $a = e + \frac{bb}{e}$, i.e., $a = \frac{ee+bb}{e}$.

Which gives

$$\cfrac{\begin{array}{c|c|c|c} +ee & +3bb & +3bb & +bb \\ ee & ee & bb & bb \\ ee & ee & ee & bb \end{array}}{eee} = +aaa$$

And

$$\cfrac{\begin{array}{c|c} -3bb & -+3bb \\ ee & bb \\ ee & ee \end{array}}{eee} = +3bba$$

$\left.\right\} = +2ccc$

100 Removing the terms that cancel and reordering the rest, it becomes $eeeee + bbbbbb = +2ccceee$.

Therefore $eeeeee - 2ccceee = -bbbbbb$.

Therefore $eeeeee - 2ccceee + cccccc = +cccccc - bbbbbb$.

Therefore $eee = ccc + \sqrt{cccccc - bbbbbb}$.

Now, on the first hypothesis in which c is greater than b, suppose that $cccccc - bbbbbb = dddddd$.

Then $eee = ccc + \sqrt{dddddd}$.

That is, $eee = ccc + ddd$, the equation stated in the first case above.

Next, on the second hypothesis in which b is equal to c, we have $cccccc - bbbbbb = 0$.

Then $eee = ccc$, the equation stated in the second case above.

And on the third hypothesis in which c is greater than b, suppose that $cccccc - bbbbbb = -dddddd$.

---

[1] The editor actually uses the following notation for cube roots here: $\sqrt{3})\sqrt{108 + 10} - \sqrt{3})\sqrt{108 - 10} = 2$.

Then $eee = ccc + \sqrt{-dddddd}$, the equation stated in the third case above – an impossible equation because $\sqrt{-dddddd}$ is inexpressible.

And in this way are achieved the required reductions of the proposed equation.

COROLLARY TO THE FIRST CASE

The resolution of the proposed equation $aaa - 3bba = +2ccc$ can also be deduced from this very reduction.

For $\overbrace{ccc + \sqrt{cccccc - bbbbbb}}^{I}$ $\overbrace{bbb}^{II}$ $\overbrace{ccc - \sqrt{cccccc - bbbbbb}}^{III}$ are in continued proportion.

Therefore so too are $\overbrace{\sqrt[3]{ccc + \sqrt{cccccc - bbbbbb}}}^{I}$ $\overset{II}{b}$

$\overbrace{\sqrt[3]{ccc - \sqrt{cccccc - bbbbbb}}}^{III}.$

But $\sqrt[3]{ccc + \sqrt{cccccc - bbbbbb}} = e$

Therefore $\sqrt[3]{ccc - \sqrt{cccccc - bbbbbb}} = \frac{bb}{e}$

Therefore $\sqrt[3]{ccc + \sqrt{cccccc - bbbbbb}} + \sqrt[3]{ccc - \sqrt{cccccc - bbbbbb}} = a$

And so the explicatory root of the proposed equation (which is what had to be expressed) is this binomial radical, itself involving binomial radicals.

Examples of solution in numbers

$40 = -6a + aaa \qquad a = \sqrt[3]{20 + \sqrt{392}} + \sqrt[3]{20 - \sqrt{392}} = 4$

$27 = -24a + aaa \qquad a = \sqrt[3]{36 + \sqrt{784}} + \sqrt[3]{36 - \sqrt{784}} = 6$

$9 = -6a + aaa \qquad a = \sqrt[3]{\frac{9}{2} + \sqrt{\frac{81}{4}}} + \sqrt[3]{\frac{9}{2} - \sqrt{\frac{81}{4}}} = 3$

101

Note 1

There is a similarity between these three cases and the three conic sections, hyperbola, parabola and ellipse, according to the presence of excess, equality or deficiency. Thus one may call the equation $aaa - 3bba = +2ccc$ by similar names (that is, hyperbolic, parabolic or elliptic). If c is greater than b, it is hyperbolic; if c is equal to b it is parabolic; and if c is less than b it is elliptic and therefore (as a type) insoluble.

Note 2

In the first two equations it sometimes happens that the binomial cube-[roots] involved in the radicals of solutions can be made clear in the same way by the binomial roots, which constitute by sum or difference a simple explicatory root of the equation. Examples of this kind of solution follow.

$$52 = \underline{-3a + aaa \qquad a = 4}$$
$$a = \underbrace{\sqrt[3]{26 + \sqrt{675}}}_{2 + \sqrt{3}} + \underbrace{\sqrt[3]{26 - \sqrt{675}}}_{2 - \sqrt{3}}$$
$$\underbrace{}_{4}$$

$$270 = \underline{+9a + aaa \qquad a = 6}$$
$$a = \underbrace{\sqrt[3]{\sqrt{18252} + 135}}_{\sqrt{12} + 3} - \underbrace{\sqrt[3]{\sqrt{18252} - 135}}_{\sqrt{12} - 3}$$
$$\underbrace{\phantom{\sqrt{12}+3 - \sqrt{12}-3}}_{6}$$

$$40 = \underline{-6a + aaa \qquad a = 4}$$
$$a = \underbrace{\sqrt[3]{20 + \sqrt{392}}}_{2 + \sqrt{2}} + \underbrace{\sqrt[3]{20 - \sqrt{292}}}_{2 - \sqrt{2}}$$
$$\underbrace{}_{4}$$

$$20 = \underline{+6a + aaa \qquad a = 2}$$
$$a = \underbrace{\sqrt[3]{\sqrt{108} + 10}}_{\sqrt{3} + 1} - \underbrace{\sqrt[3]{\sqrt{108} - 10}}_{\sqrt{3} - 1}$$
$$\underbrace{}_{2}$$

---

102 $$\sqrt{21632} = \underline{-6a + aaa \qquad a = \sqrt{32}}$$
$$a = \underbrace{\sqrt[3]{5408 + \sqrt{5400}}}_{\sqrt{8} + \sqrt{6}} + \underbrace{\sqrt[3]{5408 - \sqrt{5400}}}_{\sqrt{8} - \sqrt{6}}$$
$$\underbrace{}_{\sqrt{32}}$$

$$\sqrt{248832} = \underline{+24a + aaa \qquad a = \sqrt{48}}$$
$$a = \underbrace{\sqrt[3]{\sqrt{62720} + \sqrt{62208}}}_{\sqrt{20} + \sqrt{12}} - \underbrace{\sqrt[3]{\sqrt{62720} - \sqrt{62208}}}_{\sqrt{20} - \sqrt{12}}$$
$$\underbrace{\phantom{\sqrt{20}+\sqrt{12} - \sqrt{20}-\sqrt{12}}}_{\sqrt{48}}$$

PROBLEM 14[19]

To reduce the equation $aaaa + 4baaa = +cccc$, to the equation $eeee - 6bbee + 8bbbe = +cccc$ by putting $a = e - b$.

Let us put $e - b = a$.
Which gives $eee - 3bee + 3bbe - bbb = aaa$.

So $eeee - 4beee + 6bbee - 4bbbe + bbbb = +aaaa$ $\Big\}$
And $\quad\ + 4beee - 12bbee + 12bbbe - 4bbbb = +4baaa$ $\Big\}$ $= +cccc$

Removing the terms that cancel and reordering the rest, it becomes

$eeee - 6bbee + 8bbbe = +ccccc$
$\qquad\qquad\qquad +3bbbb,$ the sought equation, of which the root

is $e = a + b$.

And in this way is achieved the required reduction.

PROBLEM 15[20]

To reduce the equation $aaaa + 4baaa = +cccc$ to the equation

$eeee - 6bbee - 8bbbe = +cccc$
$\qquad\qquad\qquad +3bbbb$ by putting $a = e + b$, or to the equation

$eeee - 6bbee + 8bbbe = +cccc$
$\qquad\qquad\qquad +3bbbb$ by putting $a = -e + b$.

First let us put $e + b = a$.

Which gives $eee + 3bee + 3bbe + bbb = aaa$.

Hence $eeee + 3beee + 6bbee + 4bbbe + bbbb = +aaaa$ $\Big\}$
And $\qquad\quad - 4beee - 12bbee - 12bbbe - 4bbbb = -4baaa$ $\Big\}$ $= +cccc$

Removing the terms that cancel and reordering the rest, it becomes:

$eeee - 6bbee - 8bbbe = +cccc$
$\qquad\qquad\qquad +3bbbb,$

the first equation, of which the root is $e = a - b$.

Next, let us put $-e + b = a$.

Which gives $-eee + 3bee - 3bbe + bbb = aaa$.

Hence $eeee - 4beee + 6bbee - 4bbbe + bbbb = +aaaa$ $\Big\}$
And $\qquad\quad + 4beee - 12bbee + 12bbbe - 4bbbb = -4baaa$ $\Big\}$ $= +cccc$

Removing the terms that cancel and reordering the rest, it becomes:

$eeee - 6bbee + 8bbbe = +cccc$
$\qquad\qquad\qquad +3bbbb,$

the second sought equation, of which the root is $e = +b - a$.

And in this way is achieved the required reduction of the proposed equation to the sought equations.

PROBLEM 16[21]

To reduce the equation $aaaa - 4baaa = -cccc$ to the equation

$eeee - 6bbee - 8bbbe = -cccc$
$\qquad\qquad\qquad +3bbbb,$ by putting $a = e + b$, or to the equation

$eeee - 6bbee + 8bbbe = -cccc$
$\qquad\qquad\qquad +3bbbb$ by putting $a = -e + b$.

First let us put $e + b = a$.

Which gives $eee + 3bee + bbe + bbb = aaa$.

Hence $eeee + 4beee + 6bbee + 4bbbe + bbbb = +aaaa$ $\Big\}$
And $\qquad\quad - 4beee - 12bbee - 12bbbe - 4bbbb = -4baaa$ $\Big\}$ $= -cccc$

Removing the terms that cancel and reordering the rest, it becomes:

$$eeee - 6bbee - 8bbbe = -cccc$$
$$+3bbbb,$$

the first sought equation, of which the root is $e = a - b$.

Next, let us put $-e + b = a$.

Which gives $-eee + 3bee - 3bbe + bbb = aaa$.

Hence $eeee - 4beee + 6bbee - 4bbbe + bbbb = +aaaa$ ⎫
And $\qquad + 4beee - 12bbee + 12bbbe - 4bbbb = -4baaa$ ⎬ $= -cccc$
⎭

Removing the terms that cancel and reordering the rest, it becomes:

$$eeee - 6bbee + 8bbbe = -cccc$$
$$+3bbbb,$$ the second sought equation, of which

the root is $e = b - a$

And in this way is achieved the required reduction of the proposed equation to the sought equations.

PROBLEM 17

To reduce the equation $aaaa + 4baaa = -cccc$, to the equation
$$eeee - 6bbee + 8bbbe = -cccc$$
$$+3bbbb$$ by putting $a = e - b$.

Let us put $e - b = a$.

Which gives $eee - 3bee + 3bbe - bbb = aaa$.

Hence $\quad eeee - 4beee + 6bbee - 4bbbe + bbbb = +4aaaa$ ⎫
And $+ 4beee - 12bbee - 12bbbe + bbbb = +4baaa$ ⎬ $= -cccc$
⎭

Removing the terms that cancel and reordering the rest, it becomes:

$$eeee - 6bbee + 8bbbe = -cccc$$
$$+3bbbb,$$ the sought equation, of which the root

is $e = a + b$.

And in this way is achieved the required reduction.

PROBLEM 18[22]

To reduce the equation $aaaa + 4baaa + ddda = +cccc$ to the equation
$$eeee - 6bbee + 8bbbe$$
$$+ ddde = + cccc$$
$$+ 3bbbb$$ by putting $a = e - b$.
$$+ bddd$$

Let us put $e - b = a$.

Which gives $eee - 3bee + 3bbe - bbb = aaa$.

Hence $eeee - 4beee + 6bbee - 4bbbe + bbbb = +aaaa$ ⎫
And $\qquad + 4beeee - 12bbee - 12bbbe - 4bbbb = +4baaa$ ⎬ $= +cccc$
And $\qquad\qquad\qquad + ddde - bddd = +ddda$ ⎭

Removing the terms that cancel and reordering the rest, it becomes: **105**

$$eeee - 6bbee + 8bbbe$$
$$+ ddde = {+cccc \atop {+3bbbb \atop + bddd,}}$$ the sought equation, of which the root

is $e = a + b$.

And in this way is achieved the required reduction.

PROBLEM 19

To reduce the equation $aaaa - 4baaa - ddda = +cccc$ to the equation
$$eeee - 6bbee + 8bbbe$$
$$+ ddde = {+cccc \atop {+3bbbb \atop + bddd}}$$ by putting $a = -e + b$, or to the

equation
$$eeee - 6bbee - 8bbbe$$
$$- ddde = {+cccc \atop {+3bbbb \atop + bddd}}$$ by putting $a = e + b$.

Let us first put $-e + b = a$.

Which gives $-eee + 3bee - 3bbe + bbb = aaa$.

$$\left.\begin{array}{l} \text{Hence } eeee - 4beee + 6bbee - 4bbbe + bbbb = +aaaa \\ \text{And} \qquad\quad + 4beee - 12bbee + 12bbbe - 4bbbb = -4baaa \\ \text{And} \qquad\qquad\qquad\qquad + ddde - bddd = -ddda \end{array}\right\} = +cccc$$

Removing the terms that cancel and reordering the rest, it becomes:
$$eeee - 6bbee + 8bbbe$$
$$+ ddde = {+cccc \atop {+3bbbb \atop + bddd,}}$$ the first sought equation, of which the

root is $e = -a + b$.

Next, let us put $e + b = a$.

Which gives $+eee + 3bee + 3bbe + bbb = aaa$.

$$\left.\begin{array}{l} \text{Hence } eeee + 4beee + 6bbee + 4bbbe + bbbb = +aaaa \\ \text{And} \qquad\quad - 4beee - 12bbee - 12bbbe - 4bbbb = -4baaa \\ \text{And} \qquad\qquad\qquad\qquad - ddde - bddd = -ddda \end{array}\right\} = +cccc$$

Removing the terms that cancel and reordering the rest, it becomes:
$$eeee - 6bbee - 8bbbe$$
$$- ddde = {+cccc \atop {+3bbbb \atop + bddd}}$$

106

the second sought equation, of which the root is $e = a - b$.

And in this way is achieved the required reduction of the proposed equation to the sought equations.

PROBLEM 20[23]

To reduce the equation $aaaa + 4baaa + ffaa = +cccc$ to the equation

$$eeee - 6bbee + 8bbbe$$
$$+ ffee - 2bffe = +cccc$$
$$+ 3bbbb$$
$$- bbff$$

by putting $a = e - b$.

Let us put $e - b = a$.

From which $ee - 2be + be - bb = aa$

And $eee - 3bee + 3bbe - bbb = aaa$.

Hence $\left.\begin{array}{l} eeee - 4beee + 6bbee - 4bbbe + bbbb = +aaaa \\ \text{And} \quad\quad\; + 4beee - 12bbee + 12bbbe - 4bbbb = +4baaa \\ \text{And} \quad\quad\quad\quad\;\; + ffee - 2bffe + bbff = +ffaa \end{array}\right\} = +cccc$

Removing the terms that cancel and reordering the rest, it becomes:

$$eeee - 6bbee + 8bbbe$$
$$+ ffee - 2bffe = +cccc$$
$$+ 3bbbb$$
$$- bbff,$$

the sought equation, of which the root

is $e = a + b$.

And in this way is achieved the required reduction.

PROBLEM 21

To reduce the equation $aaaa - 4baaa + ffaa = +cccc$ to the equation.

$$eeee - 6bbee + 8bbbe$$
$$+ ffee - 2bffe = +cccc$$
$$+ 3bbbb$$
$$- bbff,$$

by putting $a = -e + b$, or to the

equation

$$eeee - 6bbee - 8bbbe$$
$$+ ffee + 2bbff = +cccc$$
$$+ 3bbbb$$
$$bbff$$

by putting $a = +e + b$.

Let us first put $-e + b = a$.

107 Which gives $ee - 2be + bb = aa$.

And $-eee + 3bee - 3bbe + bbb = aaa$.

Hence $\left.\begin{array}{l} eeee - 4beee + 6bbee - 4bbbe + bbbb = +aaaa \\ \text{And} \quad\quad\; + 4beee - 12bbee + 12bbbe - 4bbbb = -4baaa \\ \text{And} \quad\quad\quad\quad\;\; + ffee - 2bffe + bbff = +ffaa \end{array}\right\} = +cccc$

Removing the terms that cancel and reordering the rest, it becomes:

$$eeee - 6bbee + 8bbbe$$
$$+ ffee - 2bffe = +cccc$$
$$+ 3bbbb$$
$$- 3bbff,$$

the first sought equation, of which the root is $e = -a + b$.

Next, let us put $e + b = a$.

Which gives $ee + 2be + bb = aa$.

And $eee + 3bee + 3bbe + bbb = aaa$.

Hence $eeee + 4beee + 6bbee + 4bbbe + bbbb = +aaaa$

And $\quad\quad -4beee - 12bbee - 12bbbe - 4bbbb = -4baaa \Big\} = +cccc$

And $\quad\quad\quad\quad + ffee + 2bffe + bbff = +ffaa$

Removing the terms that cancel and reordering the rest, it becomes:

$eeee - 6bbee - 8bbbe$
$\quad + ffee + 2bffe = +cccc$
$\quad\quad\quad\quad\quad\quad + 3bbbb$ the second sought equation, of which
$\quad\quad\quad\quad\quad\quad - bbff,$

the root is $e = +a - b$.

And in this way is achieved the required reduction of the proposed equation to the sought equations.

PROBLEM 22[24]

To reduce the equation $aaaa + 4baaa + ffaa + ddda = +cccc$ to the equation

$eeee - 6bbee + 8bbbe$
$\quad + ffee - 2bffe$
$\quad\quad\quad + ddde = +cccc$
$\quad\quad\quad\quad\quad\quad + 3bbbb$ by putting $a = e - b$.
$\quad\quad\quad\quad\quad\quad - bbff$
$\quad\quad\quad\quad\quad\quad + bddd$

Let us put $e - b = a$.

Which gives $ee - 2be + bb = aa$.

And $eee - 3bee + 3bbe - bbb = aaa$.   **108**

Hence $eeee - 4beee + 6bbee - 4bbbe + bbbb = +aaaa$

And $\quad\quad +4beee - 12bbee + 12bbbe - 4bbbb = +4baaa$

And $\quad\quad\quad\quad + ffee - 2bffe + bbff = +ffaa \Big\} = +cccc$

And $\quad\quad\quad\quad\quad\quad + ddde - bddd = +ddda$

Removing the terms that cancel and reordering the rest, it becomes:

$eeee - 6bbee + 8bbbe$
$\quad + ffee - 2bffe$
$\quad\quad\quad + ddde = +cccc$
$\quad\quad\quad\quad\quad\quad + 3bbbb$
$\quad\quad\quad\quad\quad\quad - bbff$
$\quad\quad\quad\quad\quad\quad + bddd,$

the sought equation, of which the root is $e = a + b$.

And in this way is achieved the required reduction.

PROBLEM 23

To reduce the equation $aaaa - 4baaa + ffaa - ddda = +cccc$ to the equation

$eeee - 6bbee + 8bbbe$
$\quad + ffee - 2bffe$
$\quad\quad\quad + ddde = +cccc$
$\quad\quad\quad\quad\quad\quad + 3bbbb$ by putting $a = -e + b$, or to the
$\quad\quad\quad\quad\quad\quad - bbff$
$\quad\quad\quad\quad\quad\quad + bddd,$

equation

$$eeee - 6bbbe - 8bbbe$$
$$+ ffee + 2bffe$$
$$- ddde = + cccc$$
$$\quad\quad\quad + 3bbbb \quad \text{by putting } a = +e + b.$$
$$\quad\quad\quad - bbff$$
$$\quad\quad\quad + bddd$$

Let us first put $-e + b = a$.

Which gives $ee - 2be + bb = aa$.

And $-eee + 3bee - bbe + bbb = aaa$.

Hence $eeee - 4beee + 6bbee - 4bbbe + bbbb = +4aaaa$ ⎫
And $\quad\quad + 4beee - 12bbee + 12bbbe - 4bbbb = -4baaa$ ⎬ $= +cccc$
And $\quad\quad\quad + ffee - 2bffe + bbff = +ffaa$ ⎪
And $\quad\quad\quad\quad + ddde - bddd = -ddda$ ⎭

Removing the terms that cancel and reordering the rest, it becomes:

$$eeee - 6bbee + 8bbbe$$
$$+ ffee - 2bffe$$
$$+ ddde = + ccccc$$
$$\quad\quad\quad + 3bbbb \quad \text{the first sought equation, of which}$$
$$\quad\quad\quad - bbff$$
$$\quad\quad\quad + bddd,$$

109

the root is $e = -a + b$.

Next, let us put $e + b = a$.

Which gives $ee + 2be + bb = aa$.

And $eee + 3bee + 3bbe + bbb = aaa$.

Hence $eeee + 4beee + 6bbee + 4bbbe + bbbb = +aaaa$ ⎫
And $\quad\quad - 4beee - 12bbee - 12bbbe - 4bbbb = -4baaa$ ⎬ $= +cccc$
And $\quad\quad\quad + ffee + 2bffe + bbff = +ffaa$ ⎪
And $\quad\quad\quad\quad - ddde - bddd = -ddda$ ⎭

Removing the terms that cancel and reordering the rest, it becomes:

$$eeee - 6bbee - 8bbbe$$
$$+ ffee + 2bffe$$
$$- ddde = + cccc$$
$$\quad\quad\quad + 3bbbb \quad \text{the second sought equation, of which}$$
$$\quad\quad\quad - bbff$$
$$\quad\quad\quad + bddd,$$

the root is $e = a - b$.

And in this way is achieved the required reduction of the proposed equation to the sought equations.

PROBLEM 24[25]

To reduce the equation $aaaa - 4baaa + ffaa - ddda = -cccc$ to the equation

$$eeee - 6bbee - 8bbbe = - cccc$$
$$+ ffee + 2ffbe \quad + 3bbbb$$
$$- ddde \quad - ffbb \quad \text{by putting } a = e + b, \text{ or to the equation}$$
$$\quad\quad\quad + dddb,$$

$$eeee - 6bbee + 8bbbe$$
$$+ ffee \; -2ffbe$$
$$+ ddde = \begin{array}{l} -cccc \\ +3bbbb \\ -bbff \\ +bddd \end{array} \quad \text{by putting } a = -e + b.$$

The process of these reductions is similar to the preceding one.

PROBLEM 25[26]

To reduce the equation $aaaa + 4baaa - ffaa + ddda = +cccc$ to the equation **110**

$$eeee - 6bbee + 8bbbe = +cccc$$
$$- ffee \; +2bffe \quad -3bbbb$$
$$+ ddde \quad + ffbb \quad \text{by putting } a = +e - b.$$
$$+ dddb,$$

Let us put $+e - b = a$.

Which gives $ee - 2be + bb = aa$.

And $eee - 3bee + 3bbe - bbb = aaa$.

$$\left. \begin{array}{ll} \text{Hence } eeee - 4beee + & 6bbee - \; 4bbbe + \; bbbb = \; +aaaa \\ \text{And} \qquad\quad + 4beee - & 12bbee + 12bbbe - 4bbbb = +4baaa \\ \text{And} \qquad\qquad\quad - & ffee + \; 2bffe - \; bbff = \; -ffaa \\ \text{And} \qquad\qquad\qquad\quad + & ddde - \; bddd = \; +ddda \end{array} \right\} = +cccc$$

Removing the terms that cancel and reordering the rest, it becomes:

$$eeee - 6bbee + 8bbbe = +cccc$$
$$- ffee \; +2ffbe \quad +3bbbb$$
$$+ ddde \quad + ffbb \quad \text{the required equation, of which the}$$
$$+ dddb$$

root is $e = a + b$.

And in this way is achieved the required reduction of the proposed equation to the sought equation.

PROBLEM 26

To reduce the equation $aaaa + 4baaa + ffaa - ddda = +cccc$ to the equation

$$eeee - 6bbee + 8bbbe = +cccc$$
$$+ ffee \; -2ffbe \quad +3bbbb$$
$$- ddde \quad - ffbb \quad \text{by putting } a = +e - b.$$
$$- dddb$$

Let us put $+e - b = a$.

Which gives $ee - 2be + bb = aa$.

And $eee - 3bee + 3bbe - bbb = aaa$.   **111**

$$\left. \begin{array}{ll} \text{Hence } eeee - 4beee + & 6bbee - \; 4bbbe + \; bbbb = \; +aaaa \\ \text{And} \qquad\quad + 4beee - & 12bbee + 12bbbe - 4bbbb = +4baaa \\ \text{And} \qquad\qquad\quad + & ffee - \; 2ffbe + \; ffbb = \; +ffaa \\ \text{And} \qquad\qquad\qquad\quad - & ddde + \; dddb = \; -ddda \end{array} \right\} = +cccc$$

Removing the terms that cancel and reordering the rest, it becomes:

$$eeee - 6bbee + 8bbbe = +cccc$$
$$+ ffee - 2ffbe \quad + 3bbbb$$
$$- ddde \quad - ffbb$$
$$- bddd$$

the required equation, of which the root is $e = a + b$.

And in this way is achieved the required reduction of the proposed equation to the sought equation.

PROBLEM 27[27]

To reduce the equation $aaaa - 4baaa + ffaa + ddda = +cccc$ to the equation

$$eeee - 6bbee - 8bbbe = +cccc$$
$$+ ffee + 2ffbe \quad + 3bbbb$$
$$+ ddde \quad - ffbb$$
$$- dddb,$$

by putting $a = +e + b$, or to the

equation

$$eeee - 6bbee + 8bbbe$$
$$+ ffee - 2ffbe$$
$$- ddde = +cccc$$
$$+ 3bbbb$$
$$- ffbb$$
$$- dddb$$

by putting $a = -e + b$.

Let us first put $+e + b = a$.

Which gives $+ee + 2be + bb = aa$.

And $+eee + 3bee + 3bbe + bbb = aaa$.

Hence $+ eeee + 4beee + 6bbee + 4bbbe + bbbb = +aaaa$
And then $- 4beee - 12bbee - 12bbbe - 4bbbb = -4baaa$
And $+ ffee + 2bffe + bbff = +ffaa$
And $+ ddde + dddb = +ddda$

$= +cccc$

Removing the terms that cancel and reordering the rest, it becomes:

$$eeee - 6bbee - 8bbbe = +cccc$$
$$+ ffee + 2ffbe \quad + 3bbbb$$
$$+ ddde \quad - ffbb$$
$$- dddb$$

the required equation, of which the

112 root is $e = +a - b$.

Next, let us put $-e + b = a$.

Hence by a similar process it becomes

$$eeee - 6bbee + 8bbbe = +cccc$$
$$+ ffee + 2ffbe \quad + 3bbbb$$
$$- ddde \quad - ffbb$$
$$- dddb,$$

the

second required equation, of which the root is $e = -a + b$.

And in this way is achieved the required reductions of the proposed equation.

PROBLEM 28

To reduce the equation $aaaa + 4baaa - ffaa - ddda = +cccc$ to the equation

$$eeee - 6bbee + 8bbbe = +cccc$$
$$- ffee + 2ffbe \quad + 3bbbb$$
$$- ddde \quad + ffbb$$
$$- dddb$$

by putting $a = +e - b$.

Let us put $+e - b = a$.

Which gives $+ee - 2be + bb = aa$.

And $+eee - 3bee + 3bbe - bbb = aaa$.

Hence $\left.\begin{array}{l} + eeee - 4beee + 6bbee - 4bbbe + bbbb = +aaaa \\ \quad\quad\quad + 4beee - 12bbee + 12bbbe - 4bbbb = +4baaa \\ \quad\quad\quad\quad\quad\quad - ffee + 2ffbe - ffbb = -ffaa \\ \quad\quad\quad\quad\quad\quad\quad - ddde + bddd = -ddda \end{array}\right\} = +cccc$

And

And

And

Removing the terms that cancel and reordering the rest, it becomes:

$$\begin{array}{ll} eeee - 6bbee + 8bbbe = +cccc & \\ \quad - ffee \;+ ffbe \quad + 3bbbb & \\ \quad\quad\quad - ddde \quad + ffbb & \text{the required equation, of which the} \\ \quad\quad\quad\quad\quad\quad - dddb, & \end{array}$$

root is $e = +a + b$.

And in this way is achieved the required reduction of the proposed equation to the sought equation.

PROBLEM 29

To reduce the equation $aaaa - 4baaa - ffaa + ddda = +cccc$ to the equation

$$\begin{array}{ll} eeee - 6bbee - 8bbbe = +cccc & \\ \quad - ffee \;- 2ffbe \quad + 3bbbb & \\ \quad\quad\quad + ddde \quad + ffbb & \text{by putting } a = +c + b, \text{ or to the} \\ \quad\quad\quad\quad\quad - dddb & \end{array}$$

equation $\begin{array}{ll} eeee - 6bbee + 8bbbe = +cccc & \\ \quad - ffee \;+ 2ffbe \quad + 3bbbb & \\ \quad\quad\quad - ddde \quad + ffbb & \text{by putting } a = -e + b. \\ \quad\quad\quad\quad\quad - dddb & \end{array}$

113

Let us first put $+e + b = a$.

Which gives $+ee + 2be + bb = aa$.

And $+eee + 3bee + 3bbe + bbb = aaa$.

Hence $\left.\begin{array}{l} + eeee + 4beee + 6bbee + 4bbbe + bbbb = +aaaa \\ \quad\quad\quad - 4beee - 12bbee - 12bbbe - 4bbbb = -4baaa \\ \quad\quad\quad\quad\quad\quad - ffee - 2bffe - bbff = -ffaa \\ \quad\quad\quad\quad\quad\quad\quad + ddde + dddb = +ddda \end{array}\right\} = +cccc$

And then

And

And

Removing the terms that cancel and reordering the rest, it becomes:

$$\begin{array}{ll} eeee - 6bbee - 8bbbe = +cccc & \\ \quad - ffee \;- 2ffbe \quad + 3bbbb & \\ \quad\quad\quad + ddde \quad + ffbb & \text{the required equation, of which the root} \\ \quad\quad\quad\quad\quad - dddb & \end{array}$$

is $e = +a - b$.

Next, let us put $-e + b = a$.

Hence by a similar process it becomes $\begin{array}{ll} eeee - 6bbee + 8bbbe = +cccc & \\ \quad - ffee \;+ 2ffbe \quad + 3bbbb & \\ \quad\quad\quad - ddde \quad + ffbb & \\ \quad\quad\quad\quad\quad - dddb, & \end{array}$

the second equation, of which the root is $e = -a + b$.

And in this way is achieved the required reduction of the proposed equation to the sought equations.

PROBLEM 30

To reduce the equation $aaaa - 4baaa - ffaa - ddda = +cccc$ to the equation.

$$\begin{aligned} eeee - 6bbee - 8bbbe &= +cccc \\ - ffee - 2ffbe &\quad + 3bbbb \\ - ddde &\quad + ffbb \\ &\quad + dddb, \end{aligned}$$ by putting $a = +e + b$, or to the

equation $$\begin{aligned} eeee - 6bbee + 8bbbe &= +cccc \\ - ffee + 2ffbe &\quad + 3bbbb \\ + ddde &\quad + ffbb \\ &\quad + dddb \end{aligned}$$ by putting $a = -e + b$.

114 Let us first put $+e + b = a$.

Which gives $+ee + 2be + bb = aa$.

And $+eee + 3bee + 3bbe + bbb = aaa$.

$$\left.\begin{aligned} \text{Hence } eeee + 4beee + \;\;6bbee + \;\;4bbbe + \;\;bbbb &= +aaaa \\ \text{And then } \quad - 4beee - 12bbee - 12bbbe - 4bbbb &= -4baaa \\ \text{And} \qquad\qquad - \;\;ffee - \;\;2ffbe - 4ffbb &= -ffaa \\ \text{And} \qquad\qquad\qquad\quad - \;\;ddde - \;\;bddd &= -ddda \end{aligned}\right\} = +cccc$$

Removing the terms that cancel and reordering the rest, it becomes:

$$\begin{aligned} eeee - 6bbee - 8bbbe &= +cccc \\ - ffee - 2ffbe &\quad + 3bbbb \\ - ddde &\quad + ffbb \\ &\quad + dddb, \end{aligned}$$

the required equation, of which the root is $e = +a - b$.

Next, let us put $a = -e + b$.

Hence by a similar process it becomes $$\begin{aligned} eeee - 6bbee + 8bbbe &= +cccc \\ - ffee + 2ffbe &\quad + 3bbbb \\ + ddde &\quad + ffbb \\ &\quad + dddb, \end{aligned}$$ the

required equation, of which the root is $e = a + b$.

And in this way is achieved the required reduction of the proposed equation to the sought equations.

PROBLEM 31

To reduce the equation $aaaa + 4baaa - ffaa = +cccc$ to the equation

$$\begin{aligned} eeee - 6bbee + 8bbbe &= +cccc \\ - ffee + 2ffbe &\quad + 3bbbb \\ &\quad + ffbb \end{aligned}$$ by putting $a = +e - b$.

Let us put $+e - b = a$.

Which gives $+ee - 2be + bb = aa$.

And $+eee - 3bee + 3bbe - bbb = aaa$.

$$\left.\begin{aligned} \text{Hence } + eeee - 4beee + \;\;6bbee - \;\;4bbbe + \;\;bbbb &= +aaaa \\ \text{And then } \qquad + 4beee - 12bbee + 12bbbe - 4bbbb &= +4baaa \\ \text{And} \qquad\qquad\quad - \;\;ffee + \;\;2ffbe - \;\;ffbb &= -ffaa \end{aligned}\right\} = +cccc$$

Removing the terms that cancel and reordering the rest, it becomes:

$$eeee - 6bbee + 8bbbe = +cccc$$
$$- ffee + 2ffbe + 3bbbb$$
$$+ ffbb$$

115

the required equation, of which the root is $e = a + b$.

And in this way is achieved the required reduction of the proposed equation to the sought equation.

PROBLEM 32

To reduce the equation $aaaa - 4baaa - ffaa = +cccc$ to the equation.

$$eeee - 6bbee - 8bbbe = +cccc$$
$$- ffee - 2ffbe + 3bbbb$$
$$+ ffbb,$$

by putting $a = +e + b$, or to the

equation

$$eeee - 6bbee + 8bbbe = +cccc$$
$$- ffee + 2ffbe + 3bbbb$$
$$+ ffbb$$

by putting $a = -e + b$.

Let us first put $+e + b = a$.

Which gives $+ee + 2be + bb = a$.

And $+eee + 3bee + 3bbe + bbb = aaa$.

Hence $+ eeee + 4beee + 6bbee + 4bbbe + bbbb = +aaaa$

And then $- 4beee - 12bbee - 12bbbe - 4bbbb = -4baaa$ $\Big\} = +cccc$

And $- ffee - 2ffbe - ffbb = -ffaa$

Removing the terms that cancel and reordering the rest, it becomes:

$$eeee - 6bbee - 8bbbe = +cccc$$
$$- ffee - 2ffbe + 3bbbb$$
$$+ ffbb$$

the required equation, of which the

root is $e = +a - b$.

Next, let us put $-e + b = a$.

Hence by a similar process it becomes

$$eeee - 6bbee + 8bbbe$$
$$- ffee + 2ffbe = +cccc$$
$$+ 3bbbb$$
$$+ ffbb$$

the

second reduced equation, of which the root is $e = -a + b$.

And in this way is achieved the required reduction of the proposed equation to the sought equations.

PROBLEM 33

To reduce the equation $aaaa + 4baaa - ddda = +cccc$ to the equation

$$eeee - 6bbee + 8bbbe = +cccc$$
$$- ddde + 3bbbb$$
$$dddb$$

by putting $a = +e - b$.

Let us put $+e - b = a$.

116

Which gives $+ee - 2be + bb = aa$.

And $+eee - 3bee + 3bbe - bbb = aaa$.

Hence $+ eeee - 4beee + 6bbee - 4bbbe + bbbb = +aaaa$
And then $\qquad + 4beee - 12bbee + 12bbbe - 4bbbb = +4baaa$ $\Big\} = +cccc$
And $\qquad\qquad\qquad\qquad - ddde + dddb = -ddda$

Removing the terms that cancel and reordering the rest, it becomes:

$eeee - 6bbee + 8bbbe = +cccc$
$\qquad\qquad - ddde \quad + 3bbbb$ the required equation, of which the
$\qquad\qquad\qquad\qquad - dddd,$

root is $e = e + a + b$.

And in this way is achieved the required reduction of the proposed equation to the sought equation.

PROBLEM 34

To reduce the equation $aaaa - 4baaa + ddda = +cccc$ to the equation

$eeee - 6bbee - 8bbbe = +cccc$
$\qquad\quad + ddde \quad + 3bbbb$ by putting $a = e + b$.
$\qquad\qquad\quad - dddb$

Let us put $+e + b = a$.

Which gives $eee + 3bee + 3bbe + bbb = aaa$.

Hence $+ eeee + 4beee + 6bbee + 4bbbe + bbbb = +aaaa$
And $\qquad - 4beee - 12bbee - 12bbbe - 4bbbb = -4baaa$ $\Big\} = +cccc$
And $\qquad\qquad\qquad\qquad + ddde + dddb + +ddda$

Removing the terms that cancel and reordering the rest, it becomes:

$eeee - 6bbee - 8bbbe = +cccc$
$\qquad\qquad + ddde \quad + 3bbbb$ the required equation, of which the
$\qquad\qquad\qquad\qquad - bddd,$

root is $e = +a - b$.

And in this way is achieved the required reduction of the proposed equation to the sought equation.

And this brings to an end the first part of this treatise, preparatory to the numerical Exegesis, the second and principal part which follows, containing the actual practice of numerical Exegesis.

Numerical Exegesis[1]

For Solving Quadratic Equations

PROBLEM 1[2]

From the given homogeneous term of the simple quadratic equation $aa = ff$ expressed numerically, to extract by analysis the root, [which is] the value of the sought root a.

Let the numerically expressed equation be $aa = 48233025$.

And so, $48233025 = ff$.

Let us put $b + c = a$.

Then $\left.\begin{array}{c} b + c \\ b + c \end{array}\right| = 48233025$

Multiplying and ordering the homogenous elements as necessary, we get

$$\frac{\begin{array}{cc} & +2bc \\ +bb & +cc \end{array}}{\underset{Ab}{} \quad \underset{Bc}{}} = 48233025$$

Now, the pair of specious elements Ab, Bc in this equation is the canon of resolution. The analytical process must be directed by its application; the fact that it is legitimately constructed will be clear from the next Lemma.

Once the application of the canon has been done exactly as set out in the table below, with its guidance one may complete the resolution of the given homogenous term 48233025 to the root which is to be extracted from it, as follows:

Root of whole equation, to be extracted piece by piece		6	9	4	5
		.	.	.	.
Homogeneous term for resolution	ff	4 8 2 3 3 0 2 5			
Divisor		b	6		
First single root	$b = 6$	———	———		
Subtractor		bb	36		

Single root   6

Homogeneous term remaining for resolution   1 2 2 3 3 0 2 5

Divisor		$2b$	1 2 0
Ten-fold single root	$b = 60$	$2bc$	1 0 8 0
Second single root	$c = 9$	cc	8 1
Subtractor		Bc	1 1 6 1
Augmented root			6 9

Homogeneous term remaining for resolution   6 2 3 0 2 5

Divisor		$2b$	1 3 8 0
Ten-fold augmented root	$b = 690$	$2bc$	5 5 2 0
Third single root	$c = 4$	cc	1 6
Subtractor		Bc	5 5 3 6
Augmented root			6 9 4

Homogeneous term remaining for resolution   6 9 4 2 5

Divisor		$2b$	1 3 8 8 0
Ten-fold augmented root	$b = 6940$	$2bc$	6 9 4 0 0
Fourth single root	$c = 5$	cc	2 5
Subtractor		Bc	6 9 4 2 5
Root of whole equation finally extracted			6 9 4 5

The final remainder of the homogeneous term   0 0 0 0 0

And so, when the given homogeneous term 48233025 is resolved in this way, the root 6945 is extracted from it, equal to the sought root a, the required solution.

Lemma [3]

The proposed equation $aa = 48233025$ may be derived from the root 6945 (obtained by resolution) by the reverse route of composition; and thus, if we can derive the [original] equation by composition (quite obvious in this case) then, as promised above, we can check the truth both of the resolution itself and of the canon used to effect the solution.

Note 1 [4]

In those Problems which have to do with numerical Exegesis, the equation is said to be 'expressed numerically' when its given homogeneous term (if it is simple) or its given coefficients together with the homogeneous term (if it is affected) are expressed in numbers.

Note 2 [5]

119 This also should especially be noted: when forming the Canon, the two terms Ab and Bc are intended to differentiate the elements the canonical species into two

[classes]. You should know that the first term (that is, Ab) denotes those elements which pertain to the extraction of the first single root; while the second term Bc should be applied again and again in order to distinguish those elements which enable the extraction of the secondary single roots.

Moreover, you should understand this as follows: that the two terms Ab and Bc should further be distributed in four parts. A signifies the primary divisor; Ab signifies that which must be subtracted; B signifies the secondary divisor; and Bc the [other amounts] to be subtracted. And these four terms should be inserted one after another in their proper places among the elements of the canon.

But it is necessary that the signification and use of these terms be varied according to variation in the relationship or number of the elements which they are meant to signify. In the solution of simple equations, in which the divisors and primary subtractors consist of one element, it would be superfluous to apply anything beyond the terms of the canon itself (by, for instance, using the terms A, and Ab, to designate divisors and subtractors). In general, however, when the divisors and subtractors are composed of many elements, this technique is very convenient for denoting the sums of the elements (if they are of the same affection as the terms A, A, B, Bc) or their differences (if they are of contrary affections), without the trouble of writing out the almost whole canon, as one would otherwise have to do. Although this is obvious in the tables of the following examples to anyone paying the least bit of attention, it did not seem to be irrelevant to mention it here.

Problem 2[6]

From the given homogeneous term of the equation $aa + da = ff$ expressed numerically, to extract by analysis the root, [which is] the value of the sought root a.

Let the numerically expressed equation be $aa + 432a = 13584208$.

And so, $432 = d$ and $13584208 = ff$.

Let us put $b + c = a$.

Then $\dfrac{b + c}{b + c} + \dfrac{d}{b + c} = 13584208$

Multiplying and distributing the homogeneous elements into two parts, we get

$$
\begin{array}{cc}
 & +dc \\
+db & +2bc \\
+bb & +cc \\
\hline
Ab & Bc
\end{array} = 13584208
$$

Now, the pair of specious elements Ab, Bc in this equation is the canon of resolution. The analytical process must be directed by its application; the fact that it is legitimately constructed will be clear from the next Lemma.

Once the application of the canon has been done exactly as set out in the table below, with its guidance one may complete the resolution of the given homogenous term 13584208 to the root which is to be extracted from it, as follows:

120

Root of whole equation, to be extracted piece by piece				3 4 7 6
Homogeneous term for resolution ff				1 3 5 8 4 2 0 8

		d	4 3 2
Single root	$b = 3$	b	3
Divisor		A	3 4 3 2
		db	1 2 9 6
First single root	$b = 3$	bb	9
Subtractor		Ab	1 0 2 9 6
Single root			3

Homogeneous term remaining for resolution	3 2 8 8 2 0 8

		d	4 3 2
Ten-fold single root	$b = 30$	$2b$	6 0
Divisor		B	6 4 3 2
		dc	1 7 2 8
Second single root	$c = 4$	$2bc$	2 4 0
		cc	1 6
Subtractor		Bc	2 7 3 2 8
Augmented root			3 4

Homogeneous term remaining for resolution	5 5 5 4 0 8

		d	4 3 2
Ten-fold augmented root $b = 340$		$2b$	6 8 0
Divisor		B	7 2 3 2
		dc	3 0 2 4
Third single root	$c = 7$	$2bc$	4 7 6 0
		cc	4 9
Subtractor		Bc	5 1 1 1 4
Augmented root			3 4 7

The remainder for resolution	4 4 2 6 8

		d	4 3 2
Ten-fold augmented root $b = 3470$		$2b$	6 9 4 0
Divisor		B	7 3 7 2
		dc	2 5 9 2
Fourth single root	$c = 6$	$2bc$	4 1 6 4 0
		cc	3 6
Subtractor		Bc	4 4 2 6 8

Root of whole equation finally extracted	3	4	7	6

The final remainder of the homogeneous term	0 0 0 0 0

And so, when the given homogeneous term 13584208 is resolved in this way, the **121** root 3476 is extracted from it, equal to the sought root a, the required solution according to the intention of the problem.

LEMMA [7]

If from the given homogeneous term of the proposed equation $aa + 432a =$ 13584208, the root 3476, extracted by means of analysis, is equal to the sought root a and explicatory of the equation,

$$\text{then } aa = \left\lfloor \begin{matrix} 3476 \\ 3476 \end{matrix} \right. \text{ and } 432a = \left\lfloor \begin{matrix} 432 \\ 3476 \end{matrix} \right.$$

$$\text{But } \left. \begin{matrix} 3476 \\ 3476 \end{matrix} \right\rfloor = 12082576 \text{ and } \left. \begin{matrix} 432 \\ 3476 \end{matrix} \right\rfloor = 1501632$$

$$\text{And } \left. \begin{matrix} +12082576 \\ +1501632 \end{matrix} \right| = 13584208$$

Therefore $aa + 432a = $ 13584208.

But this is the proposed equation.

Thus the explication of the equation by the reverse route of composition from the root 3476 agrees; and accordingly the extracted root 3476 is equal to the sought root a, and the resolution by which the root was extracted was done correctly, and thus it follows that the canon which guided the resolution was properly formed. And this is what needed to be proved.

Case of Devolution[8]

In solving the equation $aa + da = ff$ expressed in numbers, it sometimes happens that the coefficient is extended so far back that it cannot be taken away from the homogeneous term. In such a case the coefficient should be devolved to the nearest place or third, or further if needed, until there is there is possibility of division and of beginning the work. This is made clear in the following two examples.

Example 1 of Devolution

$$\text{Equation to be solved } \begin{cases} aa + da = ff \\ aa + 75325a = 41501984 \end{cases}$$

$$\text{Canon of solution } \begin{cases} \dfrac{\begin{matrix} +db \\ +bb \end{matrix}}{Ab} & \dfrac{\begin{matrix} +dc \\ +2bc \\ +cc \end{matrix}}{Bc} \end{cases}$$

Universal root to be successively extracted $\quad\quad$ 5 $\quad$ 4 $\quad$ 7

Homogeneous term for resolution ff $\quad\quad$ 4 1 5 0 1 9 8 4

122

	d	7 5 3 2 5
	b	5
Divisor	A	7 5 8 2 5
	db	3 7 6 6 2 5
First single root $\quad b = 5$	bb	2 5
Subtractor	Ab	3 7 9 1 2 5
Single root		5

Homogeneous term remaining for resolution $\quad\quad$ 3 5 8 9 4 8 4

	d	7 5 3 2 5
Ten-fold single root $\quad b = 50$	$2b$	1 0 0
Divisor	B	7 6 3 2 5
	dc	3 0 1 3 0 0
Second single root $\quad c = 4$	$2bc$	4 0 0
	cc	1 6
Subtractor	Bc	3 0 5 4 6 0
Augmented root		5 4

Homogeneous term remaining for resolution $\quad\quad$ 5 3 4 8 8 4

	d	7 5 3 2 5
Ten-fold augmented root $b = 540$	$2b$	1 0 8 0
Divisor	B	7 6 4 0 5
	dc	5 2 7 2 7 5
Third single root $\quad c = 7$	$2bc$	7 5 6 0
	cc	4 9
Subtractor	Bc	5 3 4 8 8 4
Root of whole equation finally extracted		5 4 7

The final remainder of the homogeneous term $\quad\quad$ 0 0 0 0 0

And so, when the given homogeneous term 41501984 is resolved in this way, the root 547 is extracted from it, equal to the sought root a, which was the extraction required in this Problem.

Example 2 of Devolution

Equation to be solved $\begin{cases} aa + da = ff \\ aa + 675325a = 369701984 \end{cases}$

123

Canon of solution $\begin{cases} & +dc \\ +db & +2bc \\ \dfrac{+bb}{Ab} & \dfrac{+cc}{Bc} \end{cases}$

Root of whole equation, to be extracted piece by piece		5 4 7
Homogeneous term for resolution ff		3 6 9 7 0 1 9 8 4

	d	6 7 5 3 2 5
	b	5
Divisor	A	6 7 5 8 2 5
	db	3 3 7 6 6 2 5
First single root $\qquad b = 5$	bb	2 5
Subtractor	Ab	3 3 7 9 1 2 5
Single root		5

Homogeneous term remaining for resolution   3 1 7 8 9 4 8 4

	d	6 7 5 3 2 5
Ten-fold single root $\qquad b = 50$	$2b$	1 0 0
Divisor	B	6 7 6 3 2 5
	dc	2 7 0 1 3 0 0
Second single root $\qquad c = 4$	$2bc$	4 0 0
	cc	1 6
Subtractor	Bc	2 7 0 5 4 6 0
Augmented root		5 4

Homogeneous term remaining for resolution   4 7 3 4 8 8 4

	d	6 7 5 3 2 5
Ten-fold augmented root $b = 540$	$2b$	1 0 8 0
Divisor	B	6 7 6 4 0 5
	dc	4 7 2 7 2 7 5
Fourth single root $\qquad c = 7$	$2bc$	7 5 6 0
	cc	4 9
Subtractor	Bc	4 7 3 4 8 8 4

Root of whole equation finally extracted					5	4	7

The final remainder of the homogeneous term	0 0 0 0 0 0 0

And so, when the given homogeneous term 369701984 is resolved in this way, the root 547 is extracted from it, equal to the sought root a, which was the extraction required in this Problem.

124

PROBLEM 3[9]

From the given homogeneous term of the equation $aa - da = ff$ expressed numerically, to extract by analysis the root, [which is] the value of the sought root a.

Let the numerically expressed equation be $aa - 624a = 16305156$.

And so, $624 = d$, and $16305156 = ff$

Let us put $b + c = a$.

Then $\begin{vmatrix} b+c & -d \\ b+c & b+c \end{vmatrix} = 16305156$

Multiplying and distributing the homogeneous elements into two parts, we get

$$\begin{array}{cc} -db & -dc \\ +bb & +2bc \\ \hline \text{Ab} & +cc \\ & \hline \text{Bc} \end{array} = 16305156$$

Now, the pair of specious elements Ab, Bc in this equation is the canon of resolution. The analytical process must be directed by its application; the fact that it is legitimately constructed can be shown by the example of the Lemma above, by agreement in explication.

Once the application of the canon has been done exactly as set out in the table below, with its guidance one may complete the resolution of the given homogenous term 16305156 to the root which is to be extracted from it, as follows:

Root of whole equation, to be extracted piece by piece			4 3 6 2
Homogeneous term for resolution ff			1 6 3 0 5 1 5 6

		$-d$	$-\,6\,2\,4$
		b	4
Divisor		$\overline{A}$	$\overline{3\ 3\ 7\ 6}$
First single root	$b = 4$	$-db$	$-\,2\,4\,9\,6$
		bb	$1\ 6$
Subtractor		$\overline{Ab}$	$\overline{1\ 3\ 5\ 0\ 4}$
Single root			4
Homogeneous term remaining for resolution			$2\ 8\ 0\ 1\ 1\ 5\ 6$

		$-d$	$-$ 6 2 4
Ten-fold single root	$b = 40$	$2b$	8 0
Divisor		B	7 3 7 6
		$-dc$	$-$ 1 8 7 2
Second single root	$c = 3$	$2bc$	2 4 0
		cc	9
Subtractor		Bc	2 3 0 2 8
Augmented root			4 3

Homogeneous term remaining for resolution   4 9 8 3 5 6

		$-d$	$-$ 6 2 4
Ten-fold augmented root	$b = 430$	$2b$	8 6 0
Divisor		B	7 9 7 6
		$-dc$	$-$ 3 7 4 4
Third single root	$c = 6$	bc	5 1 6 0
		cc	3 6
Subtractor		Bc	4 8 2 1 6
Single root			4 3 6

Homogeneous term remaining for resolution   1 6 1 9 6

		$-d$	$-$ 6 2 4
Ten-fold augmented root	$b = 4360$	$2b$	8 7 2 0
Divisor		B	8 0 9 6
		$-dc$	$-$ 1 2 4 8
Fourth single root	$c = 2$	$2bc$	1 7 4 4 0
		cc	4
Subtractor		Bc	1 6 1 9 6
Root of whole equation finally extracted			4 3 6 2

Final remainder of the homogeneous term   0 0 0 0 0

And so, when the given homogeneous term 16305156 is resolved in this way, the root 4362 is extracted from it, equal to the sought root a, which was the extraction required in this Problem.

Case of Anticipation[10]

When solving the equation $aa - da = ff$ expressed in figures, it sometimes happens that the coefficient of the divisor consists of more single figures than the homogeneous term to be solved has pairs. And so, in order that a solution may take place, on the left hand side of the homogeneous term let enough zeroes be placed until it receives as many quadratic points as the coefficient has simple figures. Then, by anticipation (so to speak) the solution should begin at the first empty point. A useful aid here is that the first figure of the coefficient is either equal to or a little less than the first single root to be extracted.

Example of Anticipation

Equation to be solved $\begin{cases} aa - da = ff \\ aa - 6253a = 6254 \end{cases}$

Canon of solution $\left\{ \begin{array}{cc} \dfrac{\begin{array}{c} -db \\ +bb \end{array}}{Ab} & \dfrac{\begin{array}{c} -dc \\ +2bc \\ +cc \end{array}}{Bc} \end{array} \right.$

126

			6 2 5 4
Root of whole equation, to be extracted piece by piece			$\cdot \quad \cdot \quad \cdot \quad \cdot$
Homogeneous term for resolution ff			0 0 0 6 2 5 4
			$\cdot \quad \cdot \quad \cdot \quad \cdot$

		$-d$	− 6 2 5 3
		b	6
Divisor		$\overline{A}$	− 2 5 3
		$\overline{-db}$	− 3 7 5 1 8
First single root	$b = 6$	bb	3 6
Subtractor		$\overline{-Ab}$	− 1 5 1 8
Single root			6
			$\cdot \quad \cdot \quad \cdot \quad \cdot$
Homogeneous term remaining for resolution			1 5 2 4 2 5 4
			$\cdot \quad \cdot \quad \cdot \quad \cdot$

		$-d$	− 6 2 5 3
Ten-fold single root	$b = 50$	$2b$	1 2 0
Divisor		$\overline{B}$	5 7 4 7
		$\overline{-dc}$	− 1 2 5 0 6
Second single root	$c = 2$	$2bc$	2 4 0
		Cc	4
Subtractor		$\overline{Bc}$	1 1 8 9 4
Augmented root			6 2
			$\cdot \quad \cdot \quad \cdot \quad \cdot$
Homogeneous term remaining for resolution			3 3 4 8 5 4
			$\cdot \quad \cdot \quad \cdot \quad \cdot$

		$-d$	− 6 2 5 3
Ten-fold augmented root $b = 620$		$2b$	1 2 4 0
Divisor		$\overline{B}$	6 1 4 7
		$\overline{-dc}$	− 3 1 2 6 5
Third single root	$c = 5$	$2bc$	6 2 0 0
		cc	2 5
Subtractor		$\overline{Bc}$	3 0 9 8 5

Augmented root			6 2 5	
Homogeneous term remaining for resolution			2 5 0 0 4	

		$-d$	$-$ 6 2 5 3
Ten-fold single root	$b = 6250$	$2b$	1 2 5 0 0
Divisor		B	6 1 4 7
		$-dc$	$-$ 2 5 0 1 2
		$2bc$	5 0 0 0 0
		cc	1 6
Subtractor		Bc	2 5 0 0 4
Root of whole equation finally extracted			6 2 5 4
The final remainder of the homogeneous term			0 0 0 0 0

And so, when the given homogeneous term 6254 is resolved in this way, the root 6254 is extracted from it, equal to the sought root a, which was the extraction required in this Problem.

Case of Rectification[11]

Another thing which often happens in the equation $aa - da = ff$ expressed numerically is that the coefficient extends so far back that it makes the choice of the first single root unclear. And so it is more useful in this case to add the square of the coefficient to the homogeneous term which is to be solved. Then the first single root of that sum may be taken as the first single root of the homogeneous term which is to be solved; which root will be either equal or a little less than equal. Such a technique of extraction may be called *Epanorthosis* (to use Viète's term) or rectification.

Also in affirmed equations, if a similar case of doubt occurs in the choice of the first root, a similar remedy of rectification can be applied. In such equations, however, the first single root not of the sum but of the difference of the quadratic coefficient and the homogeneous term which is to be solved must be taken as the first single root. This also will either be equal, or a little less than equal.

Equation to be solved $\begin{cases} aa - da = ff \\ aa - 732, \quad a = 86005 \end{cases}$

Canon of solution $\begin{cases} \dfrac{\begin{array}{c} -db \\ +bb \end{array}}{Ab} & \dfrac{\begin{array}{c} -dc \\ +2bc \\ +cc \end{array}}{Bc} \end{cases}$

Example of Rectification

Given homogeneous term	ff	8 6 0 0 5
Square of the coefficient	dd	5 3 5 8 2 4
Total	$ff + dd$	6 2 1 8 2 9
First single root		8

Continued solution

Root of whole equation, to be extracted piece by piece	8 3 5
	. . .
Homogeneous term for resolution ff	8 6 0 0 5
	. . .

		$-d$	$-$ 7 3 2
	$b = 8$	$-db$	$-$ 5 8 5 6
		bb	6 4
Subtractor		Ab	5 4 4
First single root			8

Remainder of the homogeneous term to be solved   3 1 6 0 5

                                    .   .   .

128

	$b = 80$	$-d$	$-$ 7 3 2
		$2b$	1 6 0
Divisor		B	8 6 8
		$-dc$	$-$ 2 1 9 6
	$c = 3$	$2bc$	4 8 0
		cc	9
Subtractor		Bc	2 6 9 4
Augmented root			8 3

Remainder of the homogeneous term to be solved   4 6 6 5

	$b = 830$	$-d$	$-$ 7 3 2
		$2b$	1 6 6 0
Divisor		B	9 2 8
		$-dc$	$-$ 3 6 6 0
	$c = 5$	$2bc$	8 3 0 0
		cc	2 5
Subtractor		Bc	4 6 6 5

Root completely extracted		8	3	5
		$\cdot$	$\cdot$	$\cdot$
Null remainder of the homogeneous term		0 0 0 0		

PROBLEM 4[12]

From the given homogeneous term of the numerically expressed equation $-aa + da = ff$, which is explicable from two roots, to extract both roots equal to the sought root a.

The equation to be solved $\begin{cases} -aa + da = ff \\ -aa + 370a = 9261 \end{cases}$

Canon of solution $\begin{cases} \dfrac{\begin{matrix} +db \\ -bb \end{matrix}}{\text{Ab}} \quad \dfrac{\begin{matrix} +dc \\ -2bc \\ -cc \end{matrix}}{\text{Bc}} \end{cases}$

The Extraction of the Smaller Root

Root				2		7
				$\cdot$		$\cdot$
Homogeneous term for resolution				9 2 6 1		
					$\cdot$ $\cdot$	

129

			d	3 7 0
	$b = 2$		$-b$	$-$ 2
Divisor			A	3 5 0
			db	7 4 0
			$-bb$	$-$ 4
Subtractor			Ab	7 0 0
Root				2
				$\cdot$
Remainder of the homogeneous term to be solved				2 2 6 1
				$\cdot$ $\cdot$

			d	3 7 0
	$b = 20$		$-2b$	$-$ 4 0
Divisor			B	3 3 0
			dc	2 5 9 0
	$c = 7$		$-2bc$	$-$ 2 8 0
			$-cc$	$-$ 4 9
Subtractor			Bc	2 2 6 1
Root				2 7
				$\cdot$ $\cdot$
Null remainder of the homogeneous term				0 0 0 0
				$\cdot$ $\cdot$

The Extraction of the Larger Root by Anticipation

Root			3　4　3
Homogeneous term for resolution			0 9 2 6 1

		d	3 7 0
	$b = 3$	$-b$	− 3
Divisor		A	7 0
		db	1 1 1 0
		$-bb$	− 9
Subtractor		Ab	2 1 0
Root			3
Remainder of the homogeneous term to be solved			− 1 1 7 3 9

130

		d	3 7 0
	$b = 30$	$-2b$	− 6 0
Divisor		B	− 2 3 0
		dc	1 4 8 0
	$c = 4$	$-2bc$	− 2 4 0
		$-cc$	− 1 6
Subtractor		$-Bc$	− 1 0 8 0
Root			3 4
Remainder of the homogeneous term to be solved			− 9 3 9

		d	3 7 0
	$b = 340$	$-2b$	− 6 8 0
Divisor		B	− 3 1 0
		dc	1 1 1 0
	$c = 3$	$-2bc$	− 2 0 4 0
		$-cc$	− 9
Subtractor		$-Bc$	− 9 3 9
Root			3 4 3
Final remainder of the homogeneous term			0 0 0 0

And so, when a double resolution is performed in this way, two roots, 27 and 343, are extracted from the given homogeneous term 9261, as was required.

Short cut[13]

By Theorem 2 of Section 5, the sum of the two roots from which the proposed equation $-aa + da = ff$ is explicable, is equated to the given co-efficient, and the product of these is equated to the given homogeneous term, as in the numerical example $27 + 343 = 370$, and $\begin{array}{|c} 27 \\ \overline{343} \end{array} = 6261$; hence, when one is found the other is revealed without any analytical operation. This is a useful shortcut.

Note 1

It was known that the solution of quadratic equations could be treated demonstratively by the old method. But Viète did not want to lessen the esteem of his own invention by leaving out quadratics, and wished to publish his art of numerical exegesis so that it both would be general in nature and elegantly arranged by a general and complete method. Following his example, our analyst also set out the numerical exegesis of quadratic equations in his writings. And so the demands of method made it necessary to give the rules for devolution, anticipation and recti- **131** fication first in terms of quadratic equations, even though they apply to the art in general.

Note 2

Here we must note, that by multiplying by ten the roots that are used to establish the secondary divisors (as was the case in the previous tables and as will be seen in those following) the elements both of the divisor and of that which is to be subtracted uniformly end at the guide points of the equation, designated by equal degrees. And so it is straightforward and easy to order the elements. In the method hitherto recommended and used, it was wearying and difficult to order the elements because the endings of the elements were varied.

Note 3

It should also be observed that in the tables of examples the headings written in the margin are put there not for the sake of the analytic work but only to indicate the symbols of the applied canon and of the roots. During these first steps in learning the art, these were very useful. But in actual practice, where the repeated application of the canon suffices completely for direction, it will not be necessary to spell out this apparatus. And that is digression enough.

Solution of Cubic Equations

Problem 5[14]

From the given homogeneous term of the simple cubic equation $aaa = ggg$ expressed numerically, to extract by analysis the root, [which is] the value of the sought root a.

Let the numerically expressed equation be $aaa = 105689636352$

And let us put $b + c = a$.

Then $\begin{array}{c} b+c \\ b+c \\ b+c \end{array} \Bigg| = 105689636352$

Multiplying and ordering the elements, we get,

$$\underbrace{\begin{array}{l} +bbb \\ \\ \\ \end{array}}_{Ab} \underbrace{\begin{array}{l} +3bbc \\ +3bcc \\ +ccc \end{array}}_{Bc} = 105689636352$$

This is a two-part canonical equation, a guide to the analytical process, the first part Ab indicating the way to the first root, the second Bc to the second root, as will be apparent in the table below.

And so, let the extraction of the root from the given homogeneous term 105689636352 be made under the direction [of this canonical equation].

132

Universal root to be successively extracted			4	7	2	8
Homogeneous term for resolution			1 0 5 6 8 9 6 3 6 3 5 2			
Divisor	bb		1 6			
First single root	$b = 4$		——		——	
Subtractor	bbb		6 4			

** b = 4 should be be further back, lines should come between bb and bbb, and 16 and 64. **

First single root			4			
Remainder of the homogeneous term to be solved			4 1 6 8 9 6 3 6 3 5 2			
Divisor		$3bb$	4 8 0 0			
Ten-fold single root	$b = 40$	$3bbc$	3 3 6 0 0			
Second single root	$c = 7$	$3bcc$	5 8 8 0			
		ccc	3 4 3			
Subtractor		Bc	3 9 8 2 3			

Augmented root			4	7		
Remainder of the homogeneous term to be solved			1 8 6 6 6 3 6 3 5 2			
Divisor		$3bb$	6 6 2 7 0 0			
		$3bbc$	1 3 2 5 4 0 0			
Ten-fold augmented root	$b = 470$	bcc	5 6 4 0			
Third single root	$c = 2$	ccc	8			
Subtractor		Bc	1 3 3 1 0 4 8			

Augmented root			4	7	2	
Remainder of the homogeneous term to be solved			5 3 5 5 8 8 3 5 2			

Divisor	$3bb$	6 6 8 3 5 2 0 0	
	$\overline{3bbc}$	5 3 4 6 8 1 6 0 0	
Ten-fold augmented root $b = 4720$	$3bcc$	9 0 6 2 4 0	
Fourth single root $\quad c = 8$	ccc	5 1 2	
Subtractor	$\overline{Bc}$	5 3 5 5 8 8 3 5 2	

Root of whole equation finally extracted	4	7	2	8
	.	.	.	.

Final remainder of the homogeneous term	0 0 0 0 0 0 0 0 0

And so, when the given homogeneous term 105689636352 is resolved in this way, the root 4728 is extracted from it, equal to the sought root a, the required solution according to the intention of the problem.

PROBLEM 6[15]

From the given homogeneous term of the equation $aaa + daa + ffa = ggg$ expressed numerically, to extract by analysis the root, [which is] the value of the sought root a.

Let the numerically expressed equation be $aaa + 68aa + 4352a = 186394079$.   **133**

And so, $68 = d$, $4352 = ff$ and $186394079 = ggg$.

Let us put $b + c = a$.

$$\text{Then} \quad \left.\begin{array}{l} b+c \\ b+c \\ b+c \end{array}\right| \left.\begin{array}{l} +d \\ b+c \\ b+c \end{array}\right| \left.\begin{array}{l} +ff \\ b+c \end{array}\right| = 186394079$$

Multiplying the homogeneous elements we get

$$\begin{array}{l} + bbb\ + dbb\ + ffb = 186394079 \\ + 3bbc + 2dbc + ffc \\ + 3bcc + dcc \\ + ccc \end{array}$$

And distributing them into two parts gives

$$\begin{array}{l} +ffb +ffc\ +3bbc = 186394079 \\ +dbb +dcc\ +3bcc \\ \underbrace{+bbb}_{Ab}\ \underbrace{+2dbc +ccc}_{Bc} \end{array}$$

Now, the pair of specious elements Ab, Bc in this equation is the canon of resolution. The analytical process must be directed by its application; the fact that it is legitimately constructed will be clear from the next Lemma.

Once the application of the canon has been done exactly as set out in the table below, with its guidance one may complete the resolution of the given homogenous term 186394079 to the root which is to be extracted from it, as follows:

Root of whole equation, to be extracted piece by piece	5	4	7
	.	.	.

Homogeneous term for resolution ggg	1 8 6 3 9 4 0 7 9
	. . .
	. . .

		ff	0 4 3 5 2
		d	0 6 8
		bb	2 5
Divisor		A	2 6 1 1 5 2
		ffb	2 1 7 6 0
First single root	$b = 5$	dbb	1 7 0 0
		bbb	1 2 5
Subtractor		Ab	1 4 4 1 7 6 0

134

Single root	5
Remainder of the homogeneous term to be solved	4 2 2 1 8 0 7 9

		ff	4 3 5 2
		d	6 8
Ten-fold single root	$b = 50$	$2db$	6 8 0 0
		$3bb$	7 5 0 0
		$3b$	1 5 0
Divisor		B	8 3 8 0 3 2
		ffc	1 7 4 0 8
		dcc	1 0 8 8
		$2dbc$	2 7 2 0 0
Second single root	$c = 4$	$3bbc$	3 0 0 0 0
		$3bcc$	2 4 0 0
		ccc	6 4
Subtractor		Bc	3 5 4 6 6 8 8

Augmented root	5 4
Remainder of the homogeneous term to be solved	6 7 5 1 1 9 9

		ff	4 3 5 2
		d	6 8
Ten-fold augmented single root	$b = 540$	$2db$	7 3 4 4 0
		$3bb$	8 7 4 8 0 0
		$3b$	1 6 2 0
Divisor		B	9 5 4 2 8 0
		ffc	3 0 4 6 4
		dcc	3 3 3 2
		$2.dbc$	5 1 4 0 8 0
Third single root	$c = 7$	$3bbc$	6 1 2 3 6 0 0
		$3bcc$	7 9 3 8 0
		ccc	3 4 3
Subtractor		Bc	6 7 5 1 1 9 9

Root of whole equation finally extracted	5	4	7

Final remainder of the homogeneous term	0 0 0 0 0 0 0

And so, when the given homogeneous term 186394079 is resolved in this way, the root 547 is extracted from it, equal to the sought root a, the required solution.

LEMMA [16]

The proposed equation $aa + 68a + 4352a = 186394079$ may be derived from the root 547 (obtained by resolution) by the reverse route of composition; and thus, if we can derive the [original] equation by composition, we can check the truth both of the resolution itself and of the canon used to effect the solution. This is a general method of verification, which should either be actually applied to the following problems, or should be tacitly understood to be a necessary step.

Another Table of Problem 6

with a slightly different ordering of the canon[17]

Root of whole equation, to be extracted piece by piece		5	4	7

Homogeneous term for resolution ggg		1 8 6 3 9 4 0 7 9

	ff		4 3 5 2
	db		3 4 0
	bb		2 5
Divisor	A		2 8 8 3 5 2
	ffb		2 1 7 6 0
First single root $b = 5$	dbb		1 7 0 0
	bbb		1 2 5
Subtractor	Ab		1 4 4 1 7 6 0

Single root	5

Remainder of the homogeneous term to be solved	4 2 2 1 8 0 7 9

	ff	4 3 5 2
Ten-fold single root $b = 50$	$2db$	6 8 0 0
	$3bb$	7 5 0 0
Divisor	B	8 2 2 3 5 2
	ffc	1 7 4 0 8
	$2dbc$	2 7 2 0 0

	3*bbc*		3 0 0 0 0
Second single root *c* = 4	*dcc*		1 0 8 8
	3*bcc*		2 4 0 0
	ccc		6 4
Subtractor	*Bc*		3 5 4 6 6 8 8
Single root			5 4
Remainder of the homogeneous term to be solved			6 7 5 1 1 9 9

136 (margin)

	ff	4 3 5 2
Ten-fold augmented root *b* = 540	2*db*	7 3 4 4 0
	3*bb*	8 7 4 8 0 0
Divisor	*B*	9 5 2 5 9 2
	ffc	3 0 4 6 4
	2*dbc*	5 1 4 0 8 0
	3*bbc*	6 1 2 3 6 0 0
Third single root *c* = 7	*dcc*	3 3 3 2
	3*bcc*	7 9 3 8 0
	ccc	3 4 3
Subtractor	*Bc*	6 7 5 1 1 9 9
Single root completely extracted		5 4 7
Final remainder of the homogeneous term		0 0 0 0 0 0 0

Note [18]

In these two examples which pertain to the solution of the same equation you can see that a different ordering of the canon was applied to the divisors. In the first, because some approximation to the truth was made in constituting the divisor, the heterogeneous elements are taken at random to compose the divisor. Viète calls this form of division *climactic*, in which the divisor is composed out of parts of increasing degree. But in the second example, the procedure is made a little more difficult by the need to preserve the law of homogeneity. In the following examples, the climactic form is used because it is more straightforward. For, apart from the difference as stated, in climactic division the elements of the divisor are in a sense preparatory for the elements of the subtracted quantity, corresponding to them in number and order. In practice, this will be found to be a kind of a shortcut.

Problem 7 [19]

From the given homogeneous term of the equation $aaa + ffa = ggg$ expressed numerically, to extract by analysis the root, [which is] the value of the sought root a.

Let the numerically expressed equation be $aaa + 457796a = 449324752$.

And so, $45796 = ff$, and $449324752 = ggg$.

Let us put $b + c = a$.

$$\text{Then } \begin{array}{c} b+c \\ b+c \\ b+c \end{array} \left| \begin{array}{c} +ff \\ +b+c \end{array} \right. = 449324752$$

137

Multiplying the homogeneous elements, we get

$$\begin{array}{l} +bbb \ + ffb = 449324752 \\ +3bbc + ffc \\ +3bcc \\ +ccc \end{array}$$

And distributing them into two parts gives

$$\underbrace{+ffb + ffc + 3bcc}_{} = 449324752$$
$$\underbrace{+bbb}_{Ab} \ \underbrace{+ 3bbc + ccc}_{Bc}$$

Now, the pair of specious elements Ab, Bc in this equation is the canon of resolution. The analytical process must be directed by its application; the fact that it is legitimately constructed will be clear from the next Lemma.

Once the application of the canon has been done exactly as set out in the table below, with its guidance one may complete the resolution of the given homogenous term 449324752 to the root which is to be extracted from it, as follows:

Root			7	4	6
			.	.	.
Homogeneous term for resolution ggg			4 4 9 3 2 4 7 5 2		
			. . .		
		ff	4 5 7 9 6		
		bb	4 9		
Divisor		A	5 3 5 7 9 6		
		ffb	3 2 0 5 7 2		
First single root	$b = 7$	bbb	3 4 3		
Subtractor		Ab	3 7 5 0 5 7 2		
Single root			7		
			.	.	.
Remainder of the homogeneous term to be solved			7 4 2 6 7 5 5 2		
			. . .		
		ff	4 5 7 9 6		
Ten-fold single root	$b = 70$	$3bb$	1 4 7 0 0		
		$3b$	2 1 0		
Divisor		B	1 5 3 6 7 9 6		
		ffc	1 8 3 1 8 4		
		$3bbc$	5 8 8 0 0		

Second single root	$c = 4$	$3bcc$	3 3 6 0
		ccc	6 4
138 Subtractor		Bc	6 4 0 5 5 8 4
Augmented root			7 4
			.
Remainder of the homogeneous term to be solved			1 0 2 1 1 7 1 2
			. . .

		ff	4 5 7 9 6
Ten-fold augmented single root $b = 740$		$3bb$	1 6 4 2 8 0 0
		$3b$	2 2 2 0
Divisor		B	1 6 9 0 8 1 6
		ffc	2 7 4 7 7 6
		$3bbc$	9 8 5 6 8 0 0
Third single root	$c = 6$	$3bcc$	7 9 9 2 0
		ccc	2 1 6
Subtractor		Bc	1 0 2 1 1 7 1 2
Root of whole equation finally extracted			7 4 6
			. . .
Final remainder of the homogeneous term			0 0 0 0 0 0 0 0

And so, when the given homogeneous term 449324752 is resolved in this way, the root 746 is extracted from it, equal to the sought root a, the required solution.

LEMMA 12[20]

If from the given homogeneous term of the proposed equation $aaa + 45796a = 449324752$, the root 746, extracted by means of analysis, is equal to the sought root a and explicatory of the equation,

$$\text{then } aaa = \begin{array}{|c} 746 \\ 746 \\ 746 \end{array} \text{ and } 45796a = \begin{array}{|c} 45796 \\ \hline 746 \end{array}$$

$$\text{But } \begin{array}{|c} 746 \\ 746 \\ 746 \end{array} = 415160936 \text{ and } \begin{array}{|c} 45796 \\ \hline 746 \end{array} = 34163816$$

$$\text{And } \begin{array}{|c} +415160936 \\ +34163816 \end{array} = 449323752$$

Therefore $aaa + 45796a = 449324752$

But this is the proposed equation.

Thus the explication of the equation by the reverse route of composition from the root 746 agrees; and accordingly the extracted root 746 is equal to the sought root a, and the resolution by which the root was extracted was done correctly, and thus it follows that the canon which guided the resolution was properly formed. And this is what needed to be proved.

Example of Devolution[21]

Equation to be solved $\begin{cases} aaa + ffa & = ggg \\ aaa + 95400a & = 1819459 \end{cases}$

Canon of solution $\left\{ \begin{array}{l} \underbrace{+ffb}_{} \underbrace{+ffc + 3bcc}_{} \\ \underbrace{+bbb}_{Ab} \underbrace{+ 3bbc + ccc}_{Bc} \end{array} \right.$

139

Root of whole equation, to be extracted piece by piece		1 · · 9 · · ·
Homogeneous term for resolution	ggg	1 8 1 9 4 5 9 · · · ·

		ff	9 5 4 0 0	
		bb	1	
Divisor		A	9 5 4 0 0	
		ffb	9 5 4 0 0	
First single root	$b = 1$	bbb	1	
Subtractor		Ab	9 5 5 0 0	
Single root			1	
Remainder of the homogeneous term to be solved		ggg	8 6 4 4 5 9 · · · ·	

Ten-fold single root	$b = 10$	ff	9 5 4 0 0	
		$3bb$	3 0 0	
		$3b$	3 0	
Divisor		B	9 5 7 3 0	
		ffc	8 5 8 6 0 0	
		$3bbc$	2 7 0 0	
Second single root	$c = 9$	$3bcc$	2 4 3 0	
		ccc	7 2 9	
Subtractor		Bc	8 6 4 4 5 9	
Root of whole equation finally extracted			1 · · 9 ·	
Final remainder of the homogeneous term		ggg	0 0 0 0 0 0 0	

And so, the root 19 is extracted from the given homogeneous term 1819459 by analysis, equal to the sought root a, the required solution.

Example of Rectification[22]

Equation to be solved $\begin{cases} aaa + ffa & = ggg \\ aaa + 274576a & = 301163392 \end{cases}$

Canon of solution $\left\{ \begin{array}{l} \underbrace{+ffb}_{} \underbrace{+ffc + 3bcc}_{} \\ \underbrace{+bbb}_{Ab} \underbrace{+ 3bbc + ccc}_{Bc} \end{array} \right.$

140

Extraction of the First Single Root by Rectification

Given homogeneous term	ggg	3 0 1 1 6 3 3 9 2
Co-efficient of the cubic degree	$-fff$	− 1 4 3 8 7 7 8 2 4
Difference	$ggg - fff$	1 5 7 2 8 5 5 6 8
First single root	b	5

Continued Solution

Root of whole equation, to be extracted piece by piece			5 . 3 . 6 .
Homogeneous term for resolution		ggg	3 0 1 1 6 3 3 9 2 . . .
First single root	$b = 5$	ff	2 7 4 5 7 6
		ffb	1 3 7 2 8 8 0
		bbb	1 2 5
Subtractor		Ab	2 6 2 2 8 8 0
Single root			5
Remainder of the homogeneous term to be solved			3 8 8 7 5 3 9 2 . . .
Ten-fold single root	$b = 50$	ff	2 7 4 5 7 6
		$3bb$	7 5 0 0 0
		$3b$	1 5 0
Divisor		B	1 0 3 9 5 7 6
		ffc	8 2 3 7 2 8
		$3bbc$	2 2 5 0 0
Second single root	$c = 3$	$3bcc$	1 3 5 0
		ccc	2 7
Subtractor		Bc	3 2 1 1 4 2 8
Augmented root			5 . 3
Remainder of the homogeneous term to be solved			6 7 6 1 1 1 2 . . .
Ten-fold augmented root $b = 530$		ff	2 7 4 5 7 6
		$3bb$	8 4 2 7 0 0
		$3b$	1 5 9 0
Divisor		B	1 1 1 8 8 6 6
		ffc	1 6 4 7 4 5 6
		$3bbc$	5 0 5 6 2 0 0

Third single root $c = 63bcc$		5 7 2 4 0
ccc		2 1 6
Subtractor Bc		6 7 6 1 1 1 2
Root of whole equation finally extracted	5 3 6
	$\cdot$ $\cdot$ $\cdot$
Final remainder of the homogeneous term		0 0 0 0 0

141

And so, the root 536 is extracted from the given homogeneous term 305163392 by analysis, equal to the sought root a, the required solution.

PROBLEM 8[23]

From the given homogeneous term of the equation $aaa - ffa = ggg$ expressed numerically, to extract by analysis the root equal to the sought root a.

Let the numerically expressed equation be $aaa - 2648a = 91148512$.

And so, $2648 = ff$, and $91148512 = ggg$.

Let us put $b + c = a$.

Then $+ \begin{array}{c} b + c \\ b + c \\ b + c \end{array} \Bigg| - \begin{array}{c} ff \\ b + c \end{array} \Bigg| = 91148512$

Multiplying the homogeneous elements, we get

$+bbb - ffb = 91148512$
$+3bbc - ffc$
$+3bcc$
$+ccc$

And distributing them into two parts gives

$-ffb - ffc + 3bcc = 91148512$
$\underbrace{+bbb}_{Ab} \underbrace{+ 3bbc + ccc}_{Bc}$

Now, the pair of specious elements Ab, Bc in this equation is the canon of resolution. The analytical process must be directed by its application; the fact that it is legitimately constructed will be clear from the next Lemma.

Once the application of the canon has been done exactly as set out in the table below, with its guidance one may complete the resolution of the given homogenous term 91148512 to the root which is to be extracted from it, as follows:

Universal root to be extracted	4 5 2
	$\cdot$ $\cdot$ $\cdot$
Homogeneous term for resolution	9 1 1 4 8 5 1 2
	$\cdot$ $\cdot$ $\cdot$
$-ff$	$- 2 6 4 8$
bb	1 6
Divisor A	1 5 7 3 5 2

142

		$-ffb$	$- 1\ 0\ 5\ 9\ 2$
First single root	$b = 4$	bbb	6 4
Subtractor		Ab	6 2 9 4 0 8
Single root			4
Remainder of the homogeneous term to be solved			2 8 2 0 7 7 1 2

		$-ff$	$-\ 2\ 6\ 4\ 8$
Ten-fold single root	$b = 30$	$3bb$	4 8 0 0
		$3b$	1 2 0
		$+$	4 9 2 0
Divisor		B	4 8 9 3 5 2
		$-ffc$	$-\ 1\ 3\ 2\ 4\ 0$
		$3bbc$	2 4 0 0 0
Second single root	$c = 5$	$3bcc$	3 0 0 0
		ccc	1 2 5
		$+$	2 7 1 2 5
Subtractor		Bc	2 6 9 9 2 6 0
Augmented root			4 5
Remainder of the homogeneous term to be solved			1 2 1 5 1 1 2

		$-ff$	$-\ \ 2\ 6\ 4\ 8$
Ten-fold augmented root $b = 450$		$3bb$	6 0 7 5 0 0
		$3b$	1 3 5 0
		$+$	6 0 8 8 5 0
Divisor		B	6 0 6 2 0 2
		$-ffc$	$-\ \ 5\ 2\ 9\ 6$
		$3bbc$	1 2 1 5 0 0 0
Third single root	$c = 2$	$3bcc$	5 4 0 0
		ccc	8
		$+$	1 2 2 0 4 0 8
Subtractor		Bc	1 2 1 5 1 1 2
Root of whole equation finally extracted			4 5 2
Final remainder of the homogeneous term			0 0 0 0 0 0 0 0

And so, when the given homogeneous term 91148512 is resolved in this way, the root 452 is extracted from it, equal to the sought root a, the required solution according to the intention of the problem.

LEMMA[24]

If from the given homogeneous term of the proposed equation $aaa - 2648a = 91148512$, the root 452, extracted by means of analysis, is equal to the sought root a and explicatory of the equation,

then $aaa = \begin{array}{|c} 452 \\ 452 \\ 452 \end{array}$ and $2648a = \begin{array}{|c} 2468 \\ 452 \end{array}$

143

But $\begin{array}{|c} 452 \\ 452 \\ 452 \end{array} = 92345408$ and $\begin{array}{|c} 2468 \\ 452 \end{array} = 1196896$

And $\begin{array}{c} +92345408 \\ -1196896 \end{array} \Big| = 91148512$

Therefore $aaa - 2468a = 91148512$

But this is the proposed equation.

Thus the explication of the equation by the reverse route of composition from the root 452 agrees; and accordingly the extracted root 452 is equal to the sought root a, and the resolution by which the root was extracted was done correctly, and thus it follows that the canon which guided the resolution was properly formed. And this is what needed to be proved.

NOTE[25]

It should be noted here that the common signs of affection $+$ and $-$, which are applied in negatively affected equations in the linear order of the canon matched up with the other symbols, have been inserted so as to indicate individually the sums of the immediately preceding elements (when, that is, several of the same affection appear – either one of the affections or both of them). And this is done to the end that the total differences of the positive and negative elements may appear distinctly, since these are needed for establishing the divisors and subtractors. See note 2 to the first Problem.

Case of Anticipation[26]

In the equation $aaa - ffa = ggg$ expressed numerically, it sometimes happens that the coefficient of the divisors has more pairs of figures than the homogeneous term for solution has groups of three. Therefore, to make the solution possible, place enough zeroes on the left of the homogeneous term that it receives as many cubic points as the coefficient does squares; and the process of solution is begun at the first vacant point just as if by anticipation. A useful aid here is that the first quadratic root extracted from the coefficient is either equal to or a little less than the first single root to be extracted from the given homogeneous term.

Example of Anticipation

Equation to be solved $\begin{cases} aaa - ffa = ggg \\ aaa - 116620a = 352947 \end{cases}$

Canon of solution $\left| \begin{array}{cc} -ffb & -ffc + 3bcc \\ \underbrace{+bbb} & \underbrace{+3bbc + ccc} \\ Ab & Bc \end{array} \right.$

144

Root of whole equation, to be extracted piece by piece				3 4 3
Homogeneous term for resolution		ggg		0 3 5 2 9 4 7
			$-ff$	− 1 1 6 6 2 0
			bb	9
Divisor			$-A$	− 2 6 6 2 0
			$-ffb$	− 3 4 9 8 6 0
First single root		$b = 3$	bbb	2 7
Subtractor			$-Ab$	− 7 9 8 6 0
Single root				3
Remainder of the homogeneous term to be solved		ggg		8 3 3 8 9 4 7
			$-ff$	− 1 1 6 6 2 0
Ten-fold single root		$b = 30$	$3bb$	2 7 0 0
			$3b$	9 0
			$+$	2 7 9 0
			B	1 6 2 3 8 0
			$-ffc$	− 4 6 6 4 8 0
			$3bcc$	1 0 8 0 0
Second single root		$c = 4$	$3bcc$	1 4 4 0
			ccc	6 4
			$+$	1 2 3 0 4
Subtractor			Bc	7 6 3 9 2 0
Augmented root				3 4
Remainder of the homogeneous term to be solved		ggg		6 9 9 7 4 7
			$-ff$	− 1 1 6 6 2 0
Ten-fold augmented root		$b = 340$	$3bb$	3 4 6 8 0 0
			$3b$	1 0 2 0
			$+$	3 4 7 8 2 0
			B	2 3 1 2 0 0
			$-ffc$	− 3 4 9 8 6 0
			$3bcc$	1 0 4 0 4 0 0

Third single root	$c = 3$	$3bcc$				9 1 8 0	
		ccc				2 7	
		$+$				1 0 4 9 6 0 7	
Subtractor		Bc				− 6 9 9 7 4 7	
Root of whole equation finally extracted				3		4	3
				.		.	.
Final remainder of the homogeneous term	ggg				0 0 0 0 0 0	
						. . .	

And so, the root 343 is extracted from the given homogeneous term 352947 by analysis, equal to the sought root a, the required solution.

Example of Rectification

$$\begin{cases} aaa - ffa = ggg \\ aaa - 127296a = 85760000 \end{cases}$$

Canon of solution
$$\begin{vmatrix} -ffb & -ffc & +3bcc \\ +bbb & +3bbc & +ccc \end{vmatrix}$$
$$\underbrace{}_{Ab} \quad \underbrace{}_{Bc}$$

Extraction of the First Single Root by Rectification[27]

Given homogeneous term	ggg	8 5 7 6 0 0 0 0
Cubic degree of the co-efficient	fff	4 5 4 4 4 6 7 2
Total	$ggg + fff$	1 3 1 2 0 4 6 7 2
First single root	b	5
		. . .

Continued Solution

Root of whole equation, to be extracted piece by piece			5	3	6
			.	.	.
Homogeneous term for resolution		ggg	8 5 7 6 0 0 0 0		
			. . .		
		$-ff$	− 1 2 7 2 9 6		
First single root	$b = 5$	$-ffb$	− 6 3 6 4 8 0		
		bbb	1 2 5		
Subtractor		$-Ab$	6 1 3 5 2 0		
Single root			5		
			. . .		
Remainder of the homogeneous term to be solved			2 4 4 0 8 0 0 0		
			. . .		

Ten-fold single root	$b = 50$	$-ff$	$-\ 1\ 2\ 7\ 2\ 9\ 6$
		$3bb$	$7\ 5\ 0\ 0$
		$3b$	$1\ 5\ 0$
		$+$	$7\ 6\ 5\ 0$
Divisor		B	$6\ 3\ 7\ 7\ 0\ 4$
		$-ffc$	$-\ 3\ 8\ 1\ 8\ 8\ 8$
		$3bbc$	$2\ 2\ 5\ 0\ 0$
Second single root	$c = 3$	$3bcc$	$1\ 3\ 5\ 0$
		ccc	$2\ 7$
		$+$	$2\ 3\ 8\ 7\ 7$
Subtractor		Bc	$2\ 0\ 0\ 5\ 8\ 1\ 2$
Augmented root			$5 \qquad 3$
			$\cdot \qquad \cdot \qquad \cdot$
Remainder of the homogeneous term to be solved			$4\ 3\ 4\ 9\ 8\ 8\ 0$
			$\cdot \ \cdot \ \cdot$

Ten-fold augmented root $b = 530$		$-ff$	$-\ 1\ 2\ 7\ 2\ 9\ 6$
		$3bb$	$8\ 4\ 2\ 7\ 0\ 0$
		$3b$	$1\ 5\ 9\ 0$
		$+$	$8\ 4\ 4\ 2\ 9\ 0$
Divisor		B	$7\ 1\ 6\ 9\ 9\ 4$
		$-ffc$	$-\ 7\ 6\ 3\ 7\ 7\ 6$
		$3bbc$	$5\ 0\ 5\ 6\ 2\ 0\ 0$
Third single root	$c = 6$	$3bcc$	$5\ 7\ 2\ 4\ 0$
		ccc	$2\ 1\ 6$
		$+$	$5\ 1\ 1\ 3\ 6\ 5\ 6$
Subtractor		Bc	$4\ 3\ 4\ 9\ 8\ 8\ 0$
Root of whole equation finally extracted			$5 \qquad 3 \qquad 6$
			$\cdot \qquad \cdot \qquad \cdot$
Final remainder of the homogeneous term			$0\ 0\ 0\ 0\ 0$
			$\cdot \ \cdot \ \cdot$

And so, the root 536 is extracted from the given homogeneous term 85760000 by analysis, equal to the sought root a, the required solution.

PROBLEM 9[28]

From the given homogeneous term of the numerically expressed equation $-aaaa + ffa = ggg$, which is explicable from two roots, to extract by analysis both roots equal to the ambiguous root a.

Let the numerically expressed equation be $-aaaa + 52416a = 1244160$.

And so, $52416 = ff$, and $1244160 = ggg$

Let us put $b + c = a$.

Then $-\ \begin{vmatrix} b+c \\ b+c \\ b+c \end{vmatrix} +\ \begin{vmatrix} ff \\ b+c \end{vmatrix} = 1244160$

Multiplying the homogeneous elements, we get

$-bbb + ffb = 1244160$
$-3bbc + ffc$
$3bcc$
ccc

And distributing them into two parts gives

$+ffb + ffc \quad -3bcc = 1244160$
$\underbrace{-bbb}_{Ab} \quad \underbrace{-3.bbc -ccc}_{Bc}$

147

Now, the pair of specious elements Ab, Bc in this equation is the canon of resolution. The analytical process must be directed by its application; the fact that it is legitimately constructed can be demonstrated by the example of the Lemma above, by agreement in explication.

Once the application of the canon has been done exactly as set out in the table below, with its guidance one may complete the resolution of the given homogenous term 1244160 to the root which is to be extracted from it, as follows:

Extraction of the Larger Root

Root of whole equation, to be extracted piece by piece			2 . 1 . 6 .	
Homogeneous term for resolution		ggg	1 2 4 4 1 6 0 . . .	
		ff	5 2 4 1 6	
		$-bb$	− 4	
Divisor		A	1 2 4 1 6	
		ffb	1 0 4 8 3 2	
First single root	$b = 2$	$-bbb$	− 8	
Subtractor		Ab	2 4 8 3 2	
Single root			2 . . .	
Remainder of the homogeneous term to be solved			− 1 2 4 9 0 4 0 . . .	
Ten-fold single root	$b = 20$	ff	5 2 4 1 6	
		$-3bb$	− 1 2 0 0	
		$-3b$	− 6 0	
		−	− 1 2 6 0	
Divisor		$-B$	− 7 3 5 8 4	
		ffc	5 2 4 1 6	
		$-3bbc$	− 1 2 0 0	
Second single root	$c = 1$	$-3bcc$	− 6 0	
		$-ccc$	− 1	
		−	− 1 2 6 1	
Subtractor		$-Bc$	− 7 3 6 8 4	

Augmented root			2 1
			. . .
Remainder of the homogeneous term to be solved			− 5 0 2 2 0 0
			. . .

		ff	5 2 4 1 6
Ten-fold augmented root $b = 210$		$-3bb$	− 1 3 2 3 0 0
		$-3b$	− 6 3 0
		−	− 1 3 2 9 3 0
Divisor		$-B$	− 8 0 5 1 4
		ffc	3 1 4 4 9 6
		$-3bbc$	− 7 9 3 8 0 0
Third single root	$c = 6$	$-3bcc$	− 2 2 6 8 0
		$-ccc$	− 2 1 6
		−	− 8 1 6 6 9 6
Subtractor		$-Bc$	− 5 0 2 2 0 0

Root of whole equation finally extracted			2 1 6
			. . .
Final remainder of the homogeneous term			0 0 0 0 0 0
			. . .

148

Extraction of the Smaller Root by Devolution

Root of whole equation, to be extracted piece by piece			2 4
			. . .
Homogeneous term for resolution			1 2 4 4 1 6 0
			. . .

		ff	5 2 4 1 6
		$-bb$	− 4
Divisor		A	5 2 0 1 6
		ffb	1 0 4 8 3 2
First single root	$b = 2$	$-bbb$	− 8
Subtractor		Ab	1 0 4 0 3 2

Single root			2
			. . .
Remainder of the homogeneous term to be solved			2 0 3 8 4 0
			. . .

		ff	5 2 4 1 6
Ten-fold single root	$b = 20$	$-3bb$	− 1 2 0 0
		$-3b$	− 6 0
		−	− 1 2 6 0
Divisor		B	5 1 1 5 6

		ffc	2 0 9 6 4 4
		$-3bbc$	− 4 8 0 0
Second single root	$c = 1$	$-3bcc$	− 9 6 0
		$-ccc$	− 6 4
		−	− 5 8 2 4
Subtractor		$-Bc$	2 0 3 8 4 0

Root of whole equation finally extracted	2 4
	. . .
Final remainder of the homogeneous term	0 0 0 0 0 0 0
	. . .

And so, when the given homogeneous term 1244160 is resolved in this way, the two roots 216 and 24 are extracted from it, both equal to the sought root a, the required solution according to the intention of the problem.

SHORT CUT[29]

If the roots of the proposed equation $-aaa + ffa = ggg$ are put as b and c (b larger and c smaller), it will be true (by proposition 6 of Section 4) that the coefficient $ff = bb + bc + cc$.

Therefore if the greater root b is given, the equation will be $cc + bc = ff - cc$, of which the sought root is the smaller root c.

Or if the smaller root c is given, the equation will be $bb + cb = ff - cc$, of which the sought root is the greater root b.

Therefore by discovering one root of the proposed cubic equation, another can be revealed by analysis of a quadratic equation. This may be used as a short cut.

149

PROBLEM 10[30]

From the given homogeneous term of the equation $aaa - daaa = ggg$ expressed numerically, to extract by analysis the root, [which is] the value of the sought root a.

Let the numerically expressed equation be $aaa - 68a = 134454528$.

And so, $68 = d$, and $134454528 = ggg$.

Let us put $b + c = a$.

$$\text{Then} + \begin{array}{c} b+c \\ b+c \\ b+c \end{array} \bigg| - \begin{array}{c} d \\ b+c \\ b+c \end{array} \bigg| = 134454528.$$

Multiplying the homogeneous elements, we get

$+bbb - dbb = 134454528$
$+3bbc - 2dbc$
$+3bbc - dcc$
$+ccc$

And distributing them into two parts gives

$$\underbrace{\begin{matrix} -dbb \\ +bbb \end{matrix}}_{Ab} \quad \underbrace{\begin{matrix} -dcc \\ -2dbc \end{matrix} \quad \begin{matrix} +3.bbc \\ +3bcc \end{matrix} \quad +cce}_{Bc} \quad = 134454528$$

Now, the pair of specious elements Ab, Bc in this equation is the canon of resolution. The analytical process must be directed by its application; the fact that it is legitimately constructed will be clear from the next Lemma.

Once the application of the canon has been done exactly as set out in the table below, with its guidance one may complete the resolution of the given homogenous term 134454528 to the root which is to be extracted from it, as follows:

			5	3	6
Root of whole equation, to be extracted piece by piece					
			.	.	.
Homogeneous term for resolution	ggg		1 3 4 4 5 4 5 2 8		
			.	.	.

		$-d$	$-$ 6 8
		bb	2 5
Divisor		A	2 4 3 2
		$-dbb$	$-$ 1 7 0 0
First single root	$b = 4$	bbb	1 2 5
Subtractor		Ab	1 0 8 0 0
Single root			5

150

Remainder of the homogeneous term to be solved	ggg		2 6 4 5 4 5 2 8

		$-d$	$-$ 6 8
		$-2db$	$-$ 6 8 0 0
Ten-fold single root	$b = 50$	$3bb$	7 5 0 0
		$3b$	1 5 0
		$+$	7 6 5 0
		$-$	$-$ 6 8 6 8
Divisor		B	$-$ 6 9 6 3 2
		$-dcc$	$-$ 6 1 2
		$-2dbc$	$-$ 2 0 4 0 0
Second single root	$c = 3$	$3bbc$	2 2 5 0 0
		$3bcc$	1 3 5 0
		ccc	2 7
		$+$	2 3 8 7 7
		$-$	$-$ 2 1 0 1 2
Subtractor		Bc	2 1 7 7 5 8

Augmented root				5 3

Remainder of the homogeneous term to be solved

ggg			4 6 7 8 7 2 8

		$-d$	— 6 8
		$-2db$	— 7 2 0 8 0
Ten-fold single root	$b = 530$	$3bb$	8 4 2 7 0 0
		$3b$	1 5 9 0
		$+$	8 4 4 2 9 0
		$-$	– 7 2 1 4 8
Divisor		B	– 7 7 2 1 4 2
		$-dcc$	– 2 4 4 8
		$-2dbc$	– 4 3 2 4 8 0
Third single root	$c = 6$	$3bbc$	5 0 5 6 2 0 0
		$3bcc$	5 7 2 4 0
		ccc	2 1 6
		$+$	5 1 1 3 6 5 6
		$-$	– 4 3 4 9 2 8
Subtractor		Bc	4 6 7 8 7 2 8

Root of whole equation finally extracted		5 3 6
Final remainder of the homogeneous term ggg		0 0 0 0 0 0 0

And so, when the given homogeneous term 134454528 is resolved in this way, the root 452 is extracted from it, equal to the sought root a, the required solution according to the intention of the problem.

151

LEMMA 14[31]

If from the given homogeneous term of the proposed equation $aaa - 68a = 134454528$, the root 536, extracted by means of analysis, is equal to the sought root a and explicatory of the equation,

$$\text{then } aaa = \begin{array}{|c|} 536 \\ 536 \\ 536 \\ \hline \end{array} \text{ and } 68aa = \begin{array}{|c|} 68 \\ 536 \\ 536 \\ \hline \end{array}$$

$$\text{But } \begin{array}{|c|} 536 \\ 536 \\ 536 \\ \hline \end{array} = 153990656 \text{ and } \begin{array}{|c|} 68 \\ 536 \\ 536 \\ \hline \end{array} = 19536128$$

$$\text{And } \begin{array}{|c|} +153990656 \\ -19536128 \\ \hline \end{array} = 134454528$$

Therefore $aaa - 68aa = 134454528$

But this is the proposed equation.

Thus the explication of the equation by the reverse route of composition from the root 536 agrees; and accordingly the extracted root 536 is equal to the sought root a, and the resolution by which the root was extracted was done correctly, and thus it follows that the canon which guided the resolution was properly formed. And this is what needed to be proved.

For Solving Biquadratic Equations

PROBLEM 11[32]

From the given homogeneous term of the simple biquadratic equation $aaaa = hhhh$ expressed numerically, to extract by analysis the root equal to the sought root a.

Let the numerically expressed equation be $aaaaa = 19565295376$.

And let us put $b + c = a$.

$$\text{Then } \begin{vmatrix} b + c \\ b + c \\ b + c \\ b + c \end{vmatrix} = 19565295376$$

Multiplying and distributing the homogeneous elements into two parts, we get

$$\begin{aligned} +bbbb &+4bbbc = 19565295376 \\ \underbrace{}_{Ab} &+6bbcc \\ &+4bccc \\ &\underbrace{+cccc}_{Bc} \end{aligned}$$

152

Now, the pair of specious elements Ab, Bc in this equation is the canon of resolution. The analytical process must be directed by its application; the fact that it is legitimately constructed can be shown by the example of the preceding Lemmas.

Once the application of the canon has been done exactly as set out in the table below, with its guidance one may complete the resolution of the given homogenous term 19565295376 to the root which is to be extracted from it, as follows:

Root of whole equation, to be extracted piece by piece 3		7	4
		.	.
Homogeneous term for resolution	$hhhh$	1 9 5 6 5 2 9 5 3 7 6	
Divisor	bbb	2 7	
First single root	$b = 3$		
Subtractor	$bbbb$	8 1	
Single root		3	
		. . .	
Remainder of the homogeneous term to be solved		1 1 4 6 5 2 9 5 3 7 6	

		4*bbb*	1 0 8 0 0 0
Ten-fold single root	*b* = 30	6*bb*	5 4 0 0
Divisor		4*b*	1 2 0
		B	1 1 3 5 2 0
		4*bbbc*	7 5 6 0 0 0
		6*bbcc*	2 6 4 6 0 0
Second single root	*c* = 7	4*bccc*	4 1 1 6 0
		cccc	2 4 0 1
Subtractor		*Bc*	1 0 6 4 1 6 1
Augmented root			3 7

Remainder of the homogeneous term to be solved   8 2 3 6 8 5 3 7 6

		4*bbb*	2 0 2 6 1 2 0 0 0
Ten-fold augmented root *b* = 370		6*bb*	8 2 1 4 0 0
		4*b*	1 4 8 0
Divisor		*B*	2 0 3 4 3 3 8 8 0
		4*bbbc*	8 1 0 4 4 8 0 0 0
		6*bbcc*	1 3 1 4 2 4 0 0
Third single root	*c* = 4	4*bccc*	9 4 7 2 0
		cccc	2 5 6
Subtractor		*Bc*	8 2 3 6 8 5 3 7 6
Root of whole equation finally extracted			3 7 4

Final remainder of the homogeneous term   0 0 0 0 0 0 0 0 0

And so, when the given homogeneous term 19565295376 is resolved in this way, the root 374 is extracted from it, equal to the sought root *a*, the required solution according to the intention of the problem.   **153**

PROBLEM 12[33]

From the given homogeneous term of the equation $aaaa - ggga = hhhh$ expressed numerically, to extract by analysis the root, [which is] the value of the sought root *a*.

Let the numerically expressed equation be $aaaa - 426a = 2068948$.

And so, $426 = ggg$ And $2068948 = hhhh$.

Let us put $b + c = a$.

Then $+ \begin{vmatrix} b + c \\ b + c \\ b + c \\ b + c \end{vmatrix} - \dfrac{ggg}{b+c} = 2068948$

Multiplying the homogeneous elements, we get

$$+ bbbb \ \ - gggb = 206848$$
$$+ 4bbbc - gggc$$
$$+ 6bbcc$$
$$+ 4bccc$$
$$+ cccc$$

And distributing them into two parts gives

$$\underbrace{-gggb}_{} \underbrace{-gggc \ +6bbcc}_{} = 2068948$$
$$\underbrace{+bbbb}_{Ab} \underbrace{+4bbbc +4bccc + cccc}_{Bc}$$

Now, the pair of specious elements Ab, Bc in this equation is the canon of resolution. The analytical process must be directed by its application; the fact that it is legitimately constructed can be shown by the example of the preceding Lemmas.

Once the application of the canon has been done exactly as set out in the table below, with its guidance one may complete the resolution of the given homogenous term 2068948 to the root which is to be extracted from it, as follows:

Root of whole equation, to be extracted piece by piece						3		8		
								.		
Homogeneous term for resolution			*hhhh*		2 0	6 8	9 4	8		
							.	.		

			$-ggg$	$-$	4 2 6
			bbb	2 7	
Divisor			A	2 6 5 7 4	
			$-gggb$	$-$ 1 2 7 8	
First single root	$b = 5$		$bbbb$	8 1	
154 Subtractor			Ab	7 9 7 2 2	
Single root				3	

Remainder of homogeneous term to be solved		1 2 7 1 7 2 8		
			. .	

			$-ggg$	$-$	4 2 6
			$4bbb$	1 0 8 0 0 0	
Ten-fold single root	$b = 30$		$6bb$	5 4 0 0	
			$4b$	1 2 0	
			$+$	1 1 3 5 2 0	
Divisor			B	1 1 3 0 9 4	
			$-gggc$	$-$ 3 4 0 8	
			$4bbbc$	8 6 4 0 0 0	
Second single root	$c = 8$		$6bbcc$	3 4 5 6 0 0	
			$4bccc$	6 1 4 4 0	
			$cccc$	4 0 9 6	
			$+$	1 2 7 5 1 3 6	
Subtractor			Bc	1 2 7 1 7 2 8	

Root of whole equation finally extracted	3	8
	.	.
Final remainder of the homogeneous term		0 0 0 0 0 0 0

And so, when the given homogeneous term 2068948 is resolved in this way, the root 38 is extracted from it, equal to the sought root a, the required solution according to the intention of the problem.

Example of Anticipation[34]

The equation to be solved $\begin{cases} aaaa-ggga & = hhhh \\ aaaa-436023534a = 4172608 \end{cases}$

Canon of solution $\left| \begin{array}{l} -gggb-gggc\ +6bbcc \\ +bbbb+4bbbc+4bccc+cccc \end{array} \right.$

$\underbrace{}_{Ab} \quad \underbrace{}_{Bc}$

Root of whole equation, to be extracted piece by piece			3	5	2
			.	.	.
Homogeneous term for resolution		$hhhh$	0 0 4 1 7 2 6 0 8		
			. . .		

		$-ggg$	$-$ 4 3 6 0 2 3 5 4
		bbb	2 7
Divisor		$-A$	$-$ 1 6 6 0 2 3 5 4
		$-gggb$	$-$ 1 3 0 8 0 7 0 6 2
First single root	$b = 3$	$bbbb$	8 1
Subtractor		$-Ab$	$-$ 4 9 8 0 7 0 6 2

155

Single root	3
	. . .
Remainder of the homogeneous term to be solved	4 9 8 4 8 7 8 8 0 8
	. . .

		$-ggg$	$-$ 4 3 6 0 2 3 5 4
		$4bbb$	1 0 8 0 0 0
Ten-fold single root	$b = 30$	$6bb$	5 4 0 0
		$4b$	1 2 0
		$+$	1 1 3 5 2 0
Divisor		B	6 9 9 1 7 6 4 6
		$-gggc$	$-$ 2 1 8 0 1 1 7 7 0
		$4bbbc$	5 4 0 0 0 0
Second single root	$c = 5$	$6bbcc$	1 3 5 0 0 0
		$4bccc$	1 5 0 0 0
		$cccc$	6 2 5
		$+$	6 9 0 6 2 5
Subtractor		Bc	4 7 2 6 1 3 2 3 0

		Augmented root	3	5	

Augmented root   3   5

Remainder of the homogeneous term to be solved   2 5 8 7 4 6 5 0 8

		$-ggg$	$-$ 4 3 6 0 2 3 5 4
		$4bbb$	1 7 1 5 0 0 0 0 0
Ten-fold augmented root $b = 350$		$6bb$	7 3 5 0 0 0
		$4b$	1 4 0 0
		$+$	1 7 2 2 3 6 4 0 0
Divisor		B	1 2 8 6 3 4 0 5 6
		$-gggc$	$-$ 8 7 2 0 4 7 0 8
		$4bbbc$	3 4 3 0 0 0 0 0 0
Third single root	$c = 2$	$6bbcc$	2 9 4 0 0 0 0
		$4bccc$	1 1 2 0 0
		$cccc$	1 6 6
		$+$	3 4 5 9 5 1 2 1 6
Subtractor		Bc	2 5 8 7 4 6 5 0 8

Root of whole equation finally extracted   3   5   2

Final remainder of the homogeneous term   0 0 0 0 0 0 0 0 0

And so, when the given homogeneous term 4172608 is resolved in this way, the root 352 is extracted from it, equal to the sought root a, the required solution.

PROBLEM 13[35]

From the given homogeneous term of the equation $aaaa - ffaa + ggga = hhhhh$ expressed numerically, to extract by analysis the root, [which is] the value of the sought root a.

156 Let the numerically expressed equation be $aaaa - 1024aa + 6254a = 19633735875$.

And so, $1024 = ff$ and $6254 = ggg$ and $19633735875 = hhhh$.

Let us put $b + c = a$.

Then $+ \begin{vmatrix} b+c \\ b+c \\ b+c \\ b+c \end{vmatrix} - \begin{vmatrix} ff \\ b+c \\ b+c \end{vmatrix} + \dfrac{ggg}{b+c} = 19633735875$

Multiplying the homogeneous elements, we get

$+bbbb \ - ffbb \ + gggb = 19633735875$
$+4bbbc - 2ffbc + gggc$
$+6bbcc - ffcc$
$+4bccc$
$+cccc$

And distributing them into two parts gives

$$\underbrace{\begin{array}{l} +gggb +gggc \\ -ffbb -ffcc \\ +bbbb \end{array}}_{Ab} \quad \underbrace{\begin{array}{l} +4bbbc \\ +6bbcc \\ -2ffbc +4bccc + cccc \end{array}}_{Bc} = 19633735875$$

Now, the pair of specious elements Ab, Bc in this equation is the canon of resolution. The analytical process must be directed by its application; the fact that it is legitimately constructed can be demonstrated by the example of the preceding Lemmas.

Once the application of the canon has been done exactly as set out in the table below, with its guidance one may complete the resolution of the given homogenous term 19633735875 to the root which is to be extracted from it, as follows:

Root of whole equation, to be extracted piece by piece		3	7	5
Homogeneous term for resolution	$hhhh$	1 9 6 3 3 7 3 5 8 7 5		

		ggg	6 2 5 4
		$-ff$	− 1 0 2 4
		bbb	2 7
		$+$	2 7 0 0 6 2 5 4
Divisor		A	2 6 9 0 3 8 5 4
		$gggb$	1 8 7 6 2
First single root	$b = 3$	$-ffbb$	− 9 2 1 6
		$bbbb$	8 1
Subtractor		Ab	8 0 0 9 7 1 6 2
Single root			3
Remainder of the homogeneous term to be solved			1 1 6 2 4 0 1 9 6 7 5

		ggg	6 2 5 4
		$-ff$	− 1 0 2 4
		$-2ffb$	− 6 1 4 4 0
Ten-fold single root	$b = 30$	$4bbb$	1 0 8 0 0 0
		$6bb$	5 4 0 0
		$4b$	1 2 0
		$+$	1 1 3 5 2 6 2 5 4
		$-$	6 2 4 6 4
Divisor		B	1 1 2 9 0 1 6 1 4

	$gggc$		4 3 7 7 8
	$-ffcc$	−	5 0 1 7 6
	$-2ffbc$	−	4 3 0 0 8 0
Second single root	$c = 7$	$4bbbc$	7 5 6 0 0 0
	$6bbcc$		2 6 4 6 0 0
	$4bccc$		4 1 1 6 0
	$cccc$		2 4 0 1
	+		1 0 6 4 2 0 4 7 7 8
	−		4 8 0 2 5 6
Subtractor	Bc		1 0 5 9 4 0 2 2 1 8

Augmented root   3   7

Remainder of the homogeneous term to be solved   1 0 2 9 9 9 7 4 9 5

	ggg		6 2 5 4
	$-ff$	−	1 0 2 4
	$-2ffb$	−	7 5 7 7 6 0
Ten-fold augmented root $b = 370$	$4bbb$	2 0 2 6 1 2 0 0 0
	$6bb$		8 2 1 4 0 0
	$4b$		1 4 8 0
	+		2 0 3 4 4 1 1 3 4
	−		7 5 8 7 8 4
Divisor	B		2 0 2 6 8 2 3 5 0
	$gggc$		3 1 2 7 0
	$-ffcc$	−	2 5 6 0 0
	$-2ffbc$	−	3 7 8 8 8 0 0
Third single root $c = 5$	$4bbbc$	1 0 1 3 0 6 0 0 0 0
	$6bbcc$		2 0 5 3 5 0 0 0
	$4bccc$		1 8 5 0 0 0
	$cccc$		6 2 5
	+		1 0 3 3 8 1 1 8 9 5
	−		3 8 1 4 4 0 0
Subtractor	Bc		1 0 2 9 9 9 7 4 9 5

Root of whole equation finally extracted   3   7   5

Final remainder of the homogeneous term   0 0 0 0 0 0 0 0 0 0

And so, when the given homogeneous term 19633735875 is resolved in this way, the root 375 is extracted from it, equal to the sought root a, the required solution according to the intention of the problem.

PROBLEM 14[36]

From the given homogeneous term of the equation $aaaa - ffaa - ggga = hhhh$ expressed numerically, to extract by analysis the root, [which is] the value of the sought root a.

Let the numerically expressed equation be $aaaa - 1024aa - 6254a = 19629045375$.

And so, $1024aa = ff$ & $6254 = ggg$ and $19629045375 = hhhh$.

Let us put $b + c = a$.

$$\text{Then} + \begin{vmatrix} b+c \\ b+c \\ b+c \\ b+c \end{vmatrix} - \begin{vmatrix} ff \\ b+c \\ b+c \end{vmatrix} - \begin{vmatrix} ggg \\ b+c \end{vmatrix} = 19629045375$$

Multiplying the homogeneous elements, we get

$$+ bbbb \; - ffbb \; - gggb = 19629045375$$
$$+ 4bbbc - 2ffbc - gggc$$
$$+ 6bbcc - ffcc$$
$$+ 4bccc$$
$$+ cccc$$

And distributing them into two parts gives

$$\begin{array}{ll} -gggb \; -gggc & +4bbbc \\ -ffbb \; -ffcc & +6bbcc \\ \underbrace{+bbbb}_{Ab} \; \underbrace{-2ffbc +4bccc + cccc}_{Bc} \end{array} = 19629045375$$

Now, the pair of specious elements Ab, Bc in this equation is the canon of resolution. The analytical process must be directed by its application; the fact that it is legitimately constructed can be demonstrated by the example of the preceding Lemmas.

Once the application of the canon has been done exactly as set out in the table below, with its guidance one may complete the resolution of the given homogenous term 19629045375 to the root which is to be extracted from it, as follows:

159

Root of whole equation, to be extracted piece by piece	3	7	5
	.	.	.
Homogeneous term for resolution	$hhhh$	1 9 6 2 9 0 4 5 3 7 5	
		. .	.
		. . . .	

		$-ggg$	$-$ 6 2 5 4
		$-ff$	$-$ 1 0 2 4
		bbb	2 7
		$-$	$-$ 1 0 8 6 5 4
Divisor		A	2 6 8 9 1 3 4 6
		$-gggb$	$-$ 1 8 7 6 2
First single root	$b = 3$	$-ffbb$	9 2 1 6
		$bbbb$	8 1
		$-$	9 4 0 3 6 2
Subtractor		Ab	8 0 0 5 9 6 3 8

Single root   3
                                                                                 ·

Remainder of the homogeneous term to be solved   1 1 6 2 3 0 8 1 5 7 5
                                                                        ·   ·   ·
                                                                            · · ·

	$-ggg$	$-$	6 2 5 4
	$-ff$	$-$	1 0 2 4
	$-2ffb$	$-$	6 1 4 4 0
Ten-fold single root $b = 30$	$4bbb$	1 0 8 0 0 0	
	$6bb$	5 4 0 0	
	$4b$	1 2 0	
	$+$	1 1 3 5 2 0	
	$-$	6 3 0 8 9 4	
Divisor	B	1 1 2 8 8 9 1 0 6	
	$-gggc$	$-$	4 3 7 7 8
	$-ffcc$	$-$	5 0 1 7 6
	$-2ffbc$	$-$	4 3 0 0 8 0
Second single root $c = 7$	$4bbc$	7 5 6 0 0 0	
	$6bbcc$	2 6 4 6 0 0	
	$4bccc$	4 1 1 6 0	
	$cccc$	2 4 0 1	
	$+$	1 0 6 4 1 6 1	
	$-$	4 8 4 6 3 3 8	
Subtractor	Bc	1 0 5 9 3 1 4 6 6 2	

160

Augmented root   3   7
                                                                       ·   ·   ·

Remainder of the homogeneous term to be solved   1 0 2 9 9 3 4 9 5 5
                                                                        ·   ·   ·
                                                                            · · ·

	$- ggg$	$-$	6 2 5 4
	$- ff$	$-$	1 0 2 4
	$- 2ffb$	$-$	7 5 7 7 6 0
Ten-fold augmented single root $b = 370$	$4bbb$	2 0 2 6 1 2 0 0 0	
	$6bb$	8 2 1 4 0 0	
	$4b$	1 4 8 0	
	$+$	2 0 3 4 3 4 8 8 0	
	$-$	7 6 5 0 3 8	
Divisor	B	2 0 2 6 6 9 8 4 2	
	$- gggc$	$-$	3 1 2 7 0
	$- ffcc$	$-$	2 5 6 0 0
	$-2ffbc$	$-$	3 7 8 8 8 0 0
Third single root $c = 5$	$4bbc$	1 0 1 3 0 6 0 0 0 0	
	$6bbcc$	2 0 5 3 5 0 0 0	

<table>
<tr><td align="right">4bccc</td><td align="right">1 8 5 0 0 0</td></tr>
<tr><td align="right">cccc</td><td align="right">6 2 5</td></tr>
<tr><td align="center">+</td><td align="right">1 0 3 3 7 8 0 6 2 5</td></tr>
<tr><td align="center">−</td><td align="right">3 8 4 5 6 7 0</td></tr>
<tr><td>Subtractor   Bc</td><td align="right">1 0 2 9 9 3 4 9 5 5</td></tr>
</table>

Root of whole equation finally extracted	3	7	5
	.	.	.
Final remainder of the homogeneous term	0 0 0 0 0 0 0 0 0 0		

For Solving Equations of the Fifth Order

PROBLEM 15[37]

From the given homogeneous term of the simple sursolid or fifth order equation $aaaaa = lllll$ expressed numerically, to extract by analysis the root, [which is] the value of the sought root a.

Let the numerically expressed equation be $aaaaa = 15755509298176$.

And so, $lllll = 15755509298176$.

Let us put $b + c = a$.

$$\text{Then} \quad \left.\begin{array}{l} b + c \\ b + c \\ b + c \\ b + c \\ b + c \end{array}\right| = lllll$$

161

Multiplying the homogeneous elements, we get

$$\begin{array}{l} + bbbbb \quad = lllll = 15755509298176 \\ + 5bbbbc \\ + 10bbbcc \\ + 10bbccc \\ + 5bcccc \\ + ccccc \end{array}$$

And distributing them into two parts gives

$$\underbrace{+bbbbb}_{Ab} \quad \underbrace{\begin{array}{l} + 5bbbbc = 15755509298176 \\ + 10bbbcc \\ + 10bbccc \\ + 5bcccc \\ + ccccc \end{array}}_{Bc}$$

Now, the pair of specious elements Ab, Bc in this equation is the canon of resolution. The analytical process must be directed by its application; the fact that it is legitimately constructed can be demonstrated by the example of the preceding Lemmas.

Once the application of the canon has been done exactly as set out in the table below, with its guidance one may complete the resolution of the given homogenous term 15755509298176 to the root which is to be extracted from it, as follows:

Root of whole equation, to be extracted piece by piece			4	3	6
Homogeneous term for resolution			1 5 7 5 5 5 0 9 2 9 8 1 7 6		
Divisor		$bbbb$	2 5 6		
First single root	$b = 4$				
Subtractor		$bbbbb$	1 0 2 4		
Single root			4		
Remainder of the equation to be solved			5 5 1 5 5 0 9 2 9 8 1 7 6		
		$5bbbb$	1 2 8 0 0 0 0 0		
		$10bbb$	6 4 0 0 0 0		
Ten-fold single root		$10bb$	1 6 0 0 0		
	$b = 40$	$5b$	2 0 0		
Divisor		B	1 3 4 5		
		$5bbbbc$	3 8 4 0 0 0 0 0		
		$10bbbcc$	5 7 6 0 0 0 0		
Second single		$10bbccc$	4 3 2 0 0 0		
Root	$c = 3$	$5bcccc$	1 6 2 0 0		
		$ccccc$	2 4 3		
Subtractor		Bc	4 4 6 0 8 4 4 3		
Augmented root			4	3	
Remainder of the equation to be solved			1 0 5 4 6 6 4 9 9 8 1 7 6		
		$5bbbb$	1 7 0 9 4 0 0 5 0 0 0 0		
Ten-fold augmented		$10bbb$	9 9 5 0 7 0 0 0 0		
Root	$b = 430$	$10bb$	1 8 4 9 0 0 0		
		$5b$	2 1 5 0		
Divisor		B	1 7 1 9		
		$5bbbbc$	1 0 2 5 6 4 0 3 0 0 0 0 0		
		$10bbbcc$	2 8 6 2 2 5 2 0 0 0 0		
Third single root	$c = 6$	$10bbccc$	3 9 9 3 8 4 0 0 0		
		$5bcccc$	2 7 8 6 4 0 0		
		$ccccc$	7 7 7 6		
Subtractor		Bc	1 0 5 4 6 6 4 9 9 8 1 7 6		
Universal root completely extracted			4	3	6
Final remainder of the homogeneous term			0 0 0 0 0 0 0 0 0 0 0 0 0 0		

162

And so, when the given homogeneous term 15755509298176 is resolved in this way, the root 436 is extracted from it, equal to the sought root a, the required solution.

PROBLEM 16[38]

From the given homogeneous term of the equation $aaaaa - ffaaa + hhhha = lllll$ expressed numerically, to extract by analysis the root, [which is] the value of the sought root a.

Let the numerically expressed equation be $aaaaa - 57aaa + 5263a = 9000050558322$.

And so, $ff = 57$, $hhhh = 5263$ and $lllll = 900050558322$.

Let us put $b + c = a$.

$$\text{Then} + \begin{vmatrix} b+c \\ b+c \\ b+c \\ b+c \\ b+c \end{vmatrix} - \begin{vmatrix} ff \\ b+c \\ b+c \\ b+c \end{vmatrix} + \begin{vmatrix} hhhh \\ b+c \end{vmatrix} = 900050558322$$

Multiplying the homogeneous elements, we get

$$
\begin{array}{lll}
+\,bbbbb & -\,ffbbb & +\,hhhhb = 900050558322 \\
+\,5bbbbc & -\,3ffbbc & +\,hhhhc \\
+\,10bbbcc & -\,3ffbcc & \\
+\,10bbccc & -\,ffccc & \\
+\,5bcccc & & \\
+\,ccccc & &
\end{array}
$$

163

And distributing them into two parts gives

$$
\begin{array}{l}
\underbrace{+hhhhh + hhhhc}_{Ab} \quad \underbrace{-3ffbbc + 10bbccc}_{} = 900050558322 \\
-ffbbb - ffccc \quad +5bbbbc + 5bcccc \\
+bbbbb - 3ffbcc + 10bbbcc + ccccc
\end{array}
$$

$$\underbrace{}_{Ab} \qquad \underbrace{}_{Bc}$$

Now, the pair of specious elements Ab, Bc in this equation is the canon of resolution. The analytical process must be directed by its application; the fact that it is legitimately constructed can be demonstrated by the example of the preceding Lemmas.

Once the application of the canon has been done exactly as set out in the table below, with its guidance one may complete the resolution of the given homogenous term 900050558322 to the root which is to be extracted from it, as follows:

						2			4			6	
						.			.			.	
Homogeneous term for resolution	*lllll*		9	0	0	0	5	0	5	5	8	3	2	2
							.			.			.	
									.	.	.		
	hhhh								0	5	2	6	3
	$-ff$					−	0	0	0	5	7		
	bbbb					1	6						
	$+$					1	6	0	0	0	0		
Divisor	A					1	5	9	9	4			

	$hhhhb$		1 0 5 2 6
$B = 2$	$-ffbbb$		− 0 0 4 5 6
	$bbbbb$		3 2
	$+$		3 2 0 0 0 1 0 5 2 6
Subtractor	$\overline{Bc}$		3 1 9 5 4 5 0 5 2 6

2

Remainder of the homogeneous term to be solved   5 8 0 5 0 5 5 0 5 7 2 2

	$hhhh$		5 2 6 3
	$-ff$	−	5 7
	$-3ffb$	−	3 4 2 0
$b = 20$	$-3ffbb$	−	6 8 4 0 0
	$5bbbb$	8 0 0 0 0 0	
	$10bbb$	8 0 0 0 0	
	$10bb$	4 0 0 0	
	$5b$	1 0 0	
	$+$	8 8 4 1 0 0	
	$-$	7 1 8	
Divisor	$\overline{B}$	8 8 3 3 8	
	$hhhhc$		2 1 0 5 2
	$-ffccc$	−	3 6 4 8
	$-3ffbcc$	−	5 4 7 2 0
	$-3ffbbc$	−	2 7 3 6 0 0
$c = 4$	$5bbbbc$	3 2 0 0 0 0 0	
	$10bbbcc$	1 2 8 0 0 0 0	
	$10bbccc$	2 5 6 0 0 0	
	$5bcccc$	2 5 6 0 0	
	$ccccc$	1 0 2 4	
	$+$	4 7 6 2 6 2 6 1 0 5 2	
	$-$	3 3 1 9 6 8	
Subtractor	$\overline{Bc}$	4 7 5 9 3 0 6 4 2 5 2	

2   4

Remainder of the homgene to be solved   1 0 4 5 7 4 8 6 3 2 0 2

	$hhhh$		5 2 6 3
	$-ff$	−	5 7
	$-3ffb$	−	4 1 0 4 0
	$-3ffbb$	−	9 8 4 9 6 0 0
$b = 240$	$5bbbb$	1 6 5 8 8 8 0 0 0 0 0	
	$10bbb$	1 3 8 2 4 0 0 0 0	
	$10bb$	5 7 6 0 0 0	

	$5b$		1 2 0 0
	$+$	1 6 7 2 7 6 2	
	$-$	9 8 9	
Divisor	B	1 6 7 1 7	
	$hhhhc$		3 1 5 7 8
	$-ffccc$	$-$	1 2 3 1 2
	$-3ffbcc$	$-$	1 4 7 7 4 4 0
	$-3ffbbc$	$-$	5 9 0 9 7 6 0 0
$c = 6$	$5bbbbc$	9 9 5 3 2 8 0 0 0 0 0	
	$10bbbcc$	4 9 7 6 6 4 0 0 0 0	
	$10bbccc$	1 2 4 4 1 6 0 0 0	
	$5bcccc$	1 5 5 5 2 0 0	
	$ccccc$	7 7 7 6	
	$+$	1 0 4 6 3 5 4 5 0 5 5 4	
	$-$	6 0 5 8 7 3 5 2	
Subtractor	Bc	1 0 4 5 7 4 8 6 3 2 0 2	

Root of whole equation finally extracted	2	4	6

Final remainder of the homogeneous term   0 0 0 0 0 0 0 0 0 0 0

And so, when the given homogeneous term 900050558322 is resolved in this way, the root 246 is extracted from it, equal to the sought root a, the required solution.

The Practice of Approximation in the Case of Incommensurability[39]

Since it commonly happens that a root to be extracted from a given homogeneous term is irrational or inexpressible numerically (that is, when any technique of analysis is applied, at the end of the process there is always something which is left over, indicating an incomplete solution). In such a case (which is very frequent) the final remainder must be extended by progressive multiplication, adding zeroes to the right in pairs (for quadratic equations), threes (for cubics) or fours (for biquadratic) and so forth, however many one wants. Then the extracted root may, by the continuation of the same process of solution, be extended to the tenth, hundredth or thousandth part of unity – that is, to whatever degree of approximation is required. Since the process of this approximation is simple and uniform, it will be enough to illustrate it by the following examples.

Quadratic Example of Approximation

Equation to be solved $\begin{cases} aa+da = ff \\ aa+14a = 7929 \end{cases}$

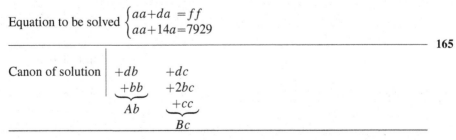

Canon of solution	$+db$	$+dc$
	$\underbrace{+bb}$	$+2bc$
	Ab	$\underbrace{+cc}$
		Bc

165

			8	2		
			·	·		
Homogeneous term for resolution			7 9 2 9			
			·	·		

		d	1 4	
		B	8	
Divisor		A	9 4	
		db	1 1 2	
	$b = 8$	bb	6 4	
Subtractor		Ab	7 5 2	

	8
	· ·
Remainder of the homogeneous term	4 0 9

		d	1 4
	$b = 80$	$2b$	1 6 0
Divisor		B	1 7 4
		dc	2 8
	$c = 2$	$2bc$	3 2 0
		cc	4
Subtractor		Bc	3 5 2

Root extracted as far as units	8 2

	· · · · · ·
Final remainder from the homogeneous term, extended	5 7 0 0 0 0 0 0
	· · · · ·

		d	1 4
	$b = 820$	$2b$	1 6 4 0
Divisor		B	1 7 8 0
		dc	4 2
	$c = 3$	$2bc$	4 9 2 0
		cc	9
Subtractor		Bc	5 3 4 9

	8 2 3
	· · · · ·
Remainder of the homogeneous term, extended	3 5 1 0 0 0 0
	· · · · ·

		d	1 4
	$b = 8230$	$2b$	1 6 4 6 0
Divisor		B	1 7 8 6 0
		dc	1 4
	$c = 1$	$2bc$	1 6 4 6 0
		cc	1
Subtractor		Bc	1 7 8 6 1

			8	2	3	1	

Remainder of the homogeneous term, extended   1 7 2 3 9 0 0

		d		1 4
	$b = 82310$	$2b$		1 6 4 6 2 0
Divisor		B		1 7 8 6 2 0
		dc		1 2 6
	$c = 9$	$2bc$		1 4 8 1 5 8 0
		cc		8 1
Subtractor		Bc		1 6 0 7 6 6 1

Root continued to thousandths   8   2   3   1   9

Remainder from the extended homogeneous term, to be discarded 1 1 6 2 3 9

And so, when the given homogeneous term 7929 is resolved in this way, the root $\frac{82319}{1000}$ or $82\frac{319}{1000}$ is obtained (that is, within one thousandth part of unity to the true root).

Cubic Example of Approximation

Equation to be solved $\begin{cases} aaa + ffa = ggg \\ aaa + 135.a = 98754 \end{cases}$

Canon of solution $\left| \begin{array}{l} +ffb+ffc +3bcc \\ +bbb +3bbc+ccc \end{array} \right.$

$\underbrace{\qquad}_{Ab} \underbrace{\qquad}_{Bc}$

Root			4	5

Homogeneous term for resolution ggg 9 8 7 5 4

		ff	1 3 5	
		bb	1 6	
Divisor		A	1 7 3 5	
		ffb	5 4 0	
	$b = 4$	bbb	6 4	
Subtractor		Ab	6 9 4 0	

                                    4

Remainder of the homogeneous term to be solved   2 9 3 5 4

	ff	1 3 5
$b = 40$	$3bb$	4 8 0 0
	$3b$	1 2 0
Divisor	B	5 0 5 5
	ffc	6 7 5
	$3bbc$	2 4 0 0 0
$c = 5$	$3bcc$	3 0 0 0
	ccc	2 2 5
Subtractor	Bc	2 7 9 0 0

Root extracted as far as units   4   5

Final remainder from the homogeneous term, extended   1 4 5 4 0 0 0 0 0 0

	ff	1 3 5
$b = 450$	$3bb$	6 0 7 5 0 0
	$3b$	1 3 5 0
Divisor	B	6 2 2
	ffc	2 7 0
$c = 2$	$3bbc$	1 2 1 5 0 0 0
	$3bcc$	5 4 0 0
	ccc	8
Subtractor	Bc	1 2 4 7 4 0 8

4   5   2

Remainder of the homogeneous term, extended   2 0 6 5 9 2 0 0 0

	ff	1 3 5
$b = 4520$	$3bb$	6 1 2 9 1 2 0 0
	$3b$	1 3 5 6 0
Divisor	B	6 2 6 5 4
	ffc	5 4 0
	$3bbc$	2 4 5 1 6 4 8 0 0
$c = 4$	$3bcc$	2 1 6 9 6 0
	ccc	6 4
Subtractor	Bc	2 5 0 7 8 1 8 2 4

Root continued to hundredths   4   5   2   4

Remainder from the extended homogeneous term, to be discarded   5 5 8 1 0 1 7 6

And so, when the given homogeneous term 98754 is resolved in this way, the root $\frac{4524}{100}$ or $45\frac{24}{100}$ is obtained (that is, within one hundredth part of unity to the true root).

Rules for Guidance

A complete and well-ordered synopsis of quadratic, cubic and biquadratic equations, and appended to each the canons for directing the analytic process. In this synopsis, the *species* of the canons are set out in four division, so that they may clearly correspond to the analytic process. Thus, for example, the two first parts, *a* and *ab*, are for the extraction of the first single root: *a* for the divisor, *ab* for the subtractor. The second two parts, *b* and *bc* are for the remaining, secondary elements: *b* for the divisor, *bc* for the subtractor.

4 Cases or Types of Quadratics

1

$aa = ff.$

A...*b*
Ab...*bb*

B... $\{\, 2b$

Bc... $\begin{cases} 2\,bc \\ cc \end{cases}$

2

$aa + da = ff.$

A $\begin{cases} d \\ b \end{cases}$

Ab $\begin{cases} db \\ bb \end{cases}$

B $\begin{cases} d \\ 2b \end{cases}$

Bc $\begin{cases} dc \\ 2\,bc \\ cc \end{cases}$

3

$aa - da = ff.$

$$A \begin{cases} -d \\ b \end{cases} \qquad B \begin{cases} -2 \\ 2b \end{cases}$$

$$Ab \begin{cases} -db \\ bb \end{cases} \qquad Bc \begin{cases} -dc \\ 2bc \\ cc \end{cases}$$

4

$-aa + da = ff.$

$$A \begin{cases} d \\ -b \end{cases} \qquad B \begin{cases} d \\ -2b \end{cases}$$

$$Ab \begin{cases} db \\ -bb \end{cases} \qquad Bc \begin{cases} dc \\ -2bc \\ -cc \end{cases}$$

---

15 Types of Cubics

1

$aaa = ggg.$

$$A \ldots bb \qquad B \begin{cases} 3bb \\ 3b \end{cases}$$
$$Ab \ldots bbb$$
$$Bc \begin{cases} 3bbc \\ 3bcc \\ ccc \end{cases}$$

2

$aaa + ffa = ggg.$

$$A \begin{cases} ff \\ bb \end{cases} \qquad B \begin{cases} ff \\ 3bb \\ 3b \end{cases}$$

$$Ab \begin{cases} ffb \\ bbb \end{cases}$$
$$Bc \begin{cases} ffc \\ 3bbc \\ 3bcc \\ ccc \end{cases}$$

3

$$aaa - ffa = ggg.$$

$$\text{A} \begin{cases} -ff \\ bb \end{cases} \qquad \text{B} \begin{cases} -ff \\ 3bb \\ 3b \end{cases}$$

$$\text{Ab} \begin{cases} -ffb \\ bbb \end{cases}$$

$$\text{Bc} \begin{cases} -ffc \\ 3bbc \\ 3bcc \\ ccc \end{cases}$$

169

4

$$-aaa + ffa = ggg.$$

$$\text{A} \begin{cases} ff \\ -bb \end{cases} \qquad \text{B} \begin{cases} ff \\ -3bb \\ -3b \end{cases}$$

$$\text{Ab} \begin{cases} ffb \\ -bbb \end{cases}$$

$$\text{Bc} \begin{cases} ffc \\ -3bbc \\ -3bcc \\ -ccc \end{cases}$$

5

$$aaa + daa = ggg.$$

$$\text{A} \begin{cases} d \\ bb \end{cases} \qquad \text{B} \begin{cases} d \\ 2db \\ 3bb \\ 3b \end{cases}$$

$$\text{Ab} \begin{cases} dbb \\ bbb \end{cases}$$

$$\text{Bc} \begin{cases} dcc \\ 2dbc \\ 3bbc \\ 3bcc \\ ccc \end{cases}$$

6

$$aaa - daa = ggg.$$

$$
A \begin{cases} -d \\ bb \end{cases} \qquad B \begin{cases} d \\ -2\,db \\ 3\,bb \\ 3\,b \end{cases}
$$

$$
Ab \begin{cases} -dbb \\ bbb \end{cases}
$$

$$
Bc \begin{cases} -dcc \\ 2\,dbc \\ 3\,bbc \\ 3\,bcc \\ ccc \end{cases}
$$

7

$$-aaa + daa = ggg.$$

$$
A \begin{cases} d \\ -bb \end{cases} \qquad B \begin{cases} d \\ 2\,db \\ -3\,bb \\ -3\,b \end{cases}
$$

$$
Ab \begin{cases} dbb \\ -bbb \end{cases}
$$

$$
Bc \begin{cases} dcc \\ 2\,dbc \\ -3\,bbc \\ -3\,bcc \\ -ccc \end{cases}
$$

8

$$aaa + daa + ffa = ggg.$$

$$
A \begin{cases} ff \\ d \\ db \end{cases} \qquad B \begin{cases} ff \\ d \\ 2\,db \\ 3\,bb \\ 3\,b \end{cases}
$$

$$
Ab \begin{cases} ffb \\ dbb \\ bbb \end{cases}
$$

$$
Bc \begin{cases} ffc \\ dcc \\ 2\,dbc \\ 3\,bbc \\ 3\,bcc \\ ccc \end{cases}
$$

9

$$aaa + daa - ffa = ggg.$$

$$A \begin{cases} -ff \\ d \\ bb \end{cases} \qquad B \begin{cases} -ff \\ d \\ 2\,db \\ 3\,bb \\ 3\,b \end{cases}$$

$$Ab \begin{cases} ffb \\ dbb \\ bbb \end{cases}$$

$$Bc \begin{cases} -ff \\ dcc \\ 2\,dbc \\ 3\,bbc \\ 3\,bcc \\ ccc \end{cases}$$

10

$$-aaa - daa + ffa = ggg.$$

$$A \begin{cases} ff \\ -d \\ -bb \end{cases} \qquad B \begin{cases} ff \\ -d \\ -2\,db \\ -3\,bb \\ -3\,b \end{cases}$$

$$Ab \begin{cases} ffb \\ -dbb \\ -bbb \end{cases}$$

$$Bc \begin{cases} ffc \\ -dcc \\ -2\,dbc \\ -3\,bbc \\ -3\,bcc \\ -ccc \end{cases}$$

170

11

$$aaa - daa + ffa = ggg.$$

$$A \begin{cases} ff \\ -d \\ bb \end{cases} \qquad B \begin{cases} ff \\ -d \\ -2\,db \\ 3\,bb \\ 3\,b \end{cases}$$

$$Ab \begin{cases} ffb \\ -dbb \\ bbb \end{cases}$$

$$Bc \begin{cases} ffc \\ -dcc \\ -2\,dbc \\ 3\,bbc \\ 3\,bcc \\ ccc \end{cases}$$

$$12$$

$$-aaa + daa - ffa = ggg.$$

$$
A \begin{cases} -ff \\ d \\ -bb \end{cases}
\qquad
B \begin{cases} -ff \\ d \\ 2\,db \\ -3\,bb \\ -3\,b \end{cases}
$$

$$
Ab \begin{cases} -ff \\ dbb \\ -bbb \end{cases}
$$

$$
Bc \begin{cases} -ffc \\ dcc \\ 2\,dbc \\ -3\,bbc \\ -3\,bcc \\ -ccc \end{cases}
$$

$$13$$

$$aaa - daa - ffa = ggg.$$

$$
A \begin{cases} -ff \\ -d \\ bb \end{cases}
\qquad
B \begin{cases} -ff \\ -d \\ -2\,db \\ 3\,bb \\ 3\,b \end{cases}
$$

$$
Ab \begin{cases} -ffb \\ -dbb \\ bbb \end{cases}
$$

$$
Bc \begin{cases} -ffc \\ -dcc \\ -2\,dbc \\ 3\,bbc \\ 3\,bcc \\ ccc \end{cases}
$$

$$14$$

$$-aaa + daa + ffa = ggg.$$

$$
A \begin{cases} ff \\ d \\ -bb \end{cases}
\qquad
B \begin{cases} ff \\ d \\ 2\,db \\ -3\,bb \\ -3\,b \end{cases}
$$

$$
Ab \begin{cases} ffb \\ dbb \\ -bbb \end{cases}
$$

$$
Bc \begin{cases} ffc \\ dcc \\ 2\,dbc \\ -3\,bbc \\ -3\,bcc \\ -ccc \end{cases}
$$

---

46 Types of Biquadratics

1

$aaaa = bbbb$

Abbb   B $\begin{cases} 4\,bbb \\ 6\,bb \\ 4\,b \end{cases}$
Abbbbb

Bc $\begin{cases} 4\,bbbc \\ 6\,bbcc \\ 4\,bccc \\ cccc \end{cases}$

2

$aaaa + ggga = bbbb$

A $\begin{cases} ggg \\ bbb \end{cases}$ B $\begin{cases} ggg \\ 4\,bbb \\ 6\,bb \\ 4\,b \end{cases}$

Ab $\begin{cases} gggb \\ bbbb \end{cases}$

Bc $\begin{cases} ggga \\ 4\,bbbc \\ 6\,bbcc \\ 4\,bccc \\ cccc \end{cases}$

171

3

$aaaa - ggga = bbbb.$

A $\begin{cases} -\,ggg \\ bbb \end{cases}$ B $\begin{cases} -\,ggg \\ 4\,bbb \\ 6\,bb \\ 4\,b \end{cases}$

Ab $\begin{cases} -\,gggb \end{cases}$

Bc $\begin{cases} -\,gggc \\ 4\,bbbc \\ 6\,bbcc \\ 4\,bccc \\ cccc \end{cases}$

4

$-aaaa + ggga = bbbb.$

$$A \begin{cases} ggg \\ -bbbb \end{cases} \qquad B \begin{cases} ggg \\ -4\,bbb \\ -6\,bb \\ -4\,b \end{cases}$$

$$Ab \begin{cases} gggb \\ -bbbb \end{cases}$$

$$Bc \begin{cases} gggc \\ -4\,bbbc \\ -6\,bccc \\ -cccc \end{cases}$$

5

$aaaa + ffaa = bbbb.$

$$A \begin{cases} ff \\ bbb \end{cases} \qquad B \begin{cases} ff \\ 2\,ffb \\ 4\,bbb \\ 6\,bb \\ 4\,b \end{cases}$$

$$Ab \begin{cases} gggb \\ bbbb \end{cases}$$

$$Bc \begin{cases} ffcc \\ 2\,ffbc \\ 4\,bbbc \\ 6\,bbcc \\ 4\,bccc \\ cccc \end{cases}$$

6

$aaaa - ffaa = bbbb.$

$$A \begin{cases} -ff \\ bbb \end{cases} \qquad B \begin{cases} -ff \\ -2\,ffb \\ 4\,bbb \\ 6\,bb \\ 4\,b \end{cases}$$

$$Ab \begin{cases} -ffbb \\ bbbb \end{cases}$$

$$Bc \begin{cases} -ffcc \\ -2\,ffbc \\ 4\,bbbc \\ 6\,bbcc \\ 4\,bccc \\ cccc \end{cases}$$

7

$-aaaa + ffaa = bbbb.$

$$A \begin{cases} ff \\ -bbb \end{cases}$$

$$Ab \begin{cases} ff \\ -bbbb \end{cases}$$

$$B \begin{cases} ff \\ 2\,ffb \\ -4\,bbb \\ -6\,bb \\ -4\,b \end{cases}$$

$$Bc \begin{cases} ffcc \\ 2\,ffbc \\ -4\,bbbc \\ -6\,bbcc \\ -4\,bccc \\ -cccc \end{cases}$$

8

$aaaa + daaa = bbbb.$

$$A \begin{cases} d \\ bbb \end{cases}$$

$$Ab \begin{cases} dbbb \\ bbbb \end{cases}$$

$$B \begin{cases} d \\ 3\,db \\ 3\,dbb \\ 4\,bbb \\ 6\,bb \\ 4\,b \end{cases}$$

$$Bc \begin{cases} dccc \\ 3\,dbcc \\ 3\,dbbc \\ 4\,bbbc \\ 6\,bbcc \\ 4\,bccc \\ cccc \end{cases}$$

172

9

$aaaa - daaa = bbbb.$

$$A \begin{cases} -d \\ bbb \end{cases}$$

$$Ab \begin{cases} -dbbb \\ bbbb \end{cases}$$

$$B \begin{cases} -d \\ -3\,db \\ -3\,dbbb \\ 4\,bbb \\ 6\,bb \\ 4\,b \end{cases}$$

$$Bc \begin{cases} -dccc \\ -3\,dbcc \\ -3\,dbbc \\ 4\,bbbc \\ 6\,bbcc \\ 4\,bccc \\ cccc \end{cases}$$

10

$-aaaa + daaa = bbbb$.

$$A \begin{cases} d \\ -bbb \end{cases} \qquad B \begin{cases} d \\ 3\,db \\ 3\,dbb \\ -4\,bbb \\ -6\,bb \\ -4\,b \end{cases}$$

$$Ab \begin{cases} dbbb \\ -bbbb \end{cases}$$

$$Bc \begin{cases} dccc \\ 3\,dbcc \\ 3\,dbbc \\ -4\,bbbc \\ -6\,bbcc \\ -4\,bccc \\ -cccc \end{cases}$$

11

$aaaa + ffaa + ggga = bbbb$.

$$A \begin{cases} ggg \\ ff \\ bbb \end{cases} \qquad B \begin{cases} ggg \\ ff \\ 2\,ffb \\ 4\,bbb \\ 6\,bb \\ 4\,b \end{cases}$$

$$Ab \begin{cases} gggb \\ ffbb \\ bbbb \end{cases}$$

$$Bc \begin{cases} gggc \\ ffcc \\ 2\,ffbc \\ 4\,bbbc \\ 6\,bbcc \\ 4\,bccc \\ cccc \end{cases}$$

12

$aaaa + ffaa - ggga = bbbb$.

$$A \begin{cases} -ggg \\ ff \\ bbb \end{cases} \qquad B \begin{cases} -ggg \\ ff \\ 2\,ffb \\ 4\,bbb \\ 6\,bb \\ 4\,b \end{cases}$$

$$Ab \begin{cases} -gggb \\ ffbb \\ bbbb \end{cases}$$

$$Bc \begin{cases} -gggc \\ ffcc \\ 2\,ffbc \\ 4\,bbbc \\ 6\,bbcc \\ 4\,bccc \\ cccc \end{cases}$$

13

$-aaaa - ffaa + ggga = bbbb.$

$$
A \begin{cases} ggg \\ -ff \\ -bbb \end{cases} \quad
B \begin{cases} ggg \\ -ff \\ -2fab \\ -4bbb \\ -6bb \\ -4b \end{cases}
$$

$$
Ab \begin{cases} gggb \\ -ffbb \\ -bbbb \end{cases}
$$

$$
Bc \begin{cases} gggc \\ -ffcc \\ -2ffbc \\ -4bbbc \\ -6bbcc \\ -4bccc \\ -cccc \end{cases}
$$

14

$aaaa - ffaa + ggga = bbbb.$

$$
A \begin{cases} ggg \\ -ff \\ bbb \end{cases} \quad
B \begin{cases} ggg \\ -ff \\ -2ffb \\ 4bbb \\ 6bb \\ 4b \end{cases}
$$

$$
Ab \begin{cases} gggb \\ -ffbb \\ bbbb \end{cases}
$$

$$
Bc \begin{cases} gggc \\ -ffcc \\ -2ffbc \\ 4bbbc \\ 4bccc \\ cccc \end{cases}
$$

173

15

$-aaaa + ffaa - ggga = bbbb.$

$$
A \begin{cases} -ggg \\ ff \\ -bbb \end{cases} \quad
B \begin{cases} -ggg \\ ff \\ 2ffb \\ -4bbb \\ -6bb \\ -4b \end{cases}
$$

$$
Ab \begin{cases} -gggb \\ ffbb \\ -bbbb \end{cases}
$$

$$
Bc \begin{cases} -gggc \\ ffc \\ 2ffbc \\ -4bbbc \\ -6bbcc \\ -4bccc \\ -cccc \end{cases}
$$

16

$$aaaa - ffaa - ggga = bbbb.$$

$$
A \begin{cases} -ggg \\ -ff \\ bbbb \end{cases}
\qquad
B \begin{cases} -ggg \\ -ff \\ -2ffb \\ 4bbb \\ 6bb \\ 4b \end{cases}
$$

$$
Ab \begin{cases} -gggb \\ -ffbb \\ bbbb \end{cases}
$$

$$
Bc \begin{cases} -gggc \\ -ffcc \\ -2ffbc \\ 4bbbc \\ 6bbcc \\ 4bccc \\ cccc \end{cases}
$$

17

$$-aaaa + ffaa + ggga = bbbb.$$

$$
A \begin{cases} ggg \\ ff \\ -bbb \end{cases}
\qquad
B \begin{cases} ggg \\ ff \\ 2ffb \\ -4bbb \\ -6bb \\ -4b \end{cases}
$$

$$
Ab \begin{cases} gggb \\ ffbb \\ -bbbb \end{cases}
$$

$$
Bc \begin{cases} gggc \\ ffcc \\ 2ffbc \\ -4bbbc \\ -6bbcc \\ -4bccc \end{cases}
$$

18

$$aaaa + daaa + ggga = bbbb.$$

$$
A \begin{cases} ggg \\ d \\ bbb \end{cases}
\qquad
B \begin{cases} ggg \\ d \\ 3db \\ 3dbb \\ 4bbb \\ 6bb \\ 4b \end{cases}
$$

$$
Ab \begin{cases} gggb \\ dbbb \\ bbbb \end{cases}
$$

$$
Bc \begin{cases} gggc \\ dccc \\ 3dbcc \\ 3dbbc \\ 4bbbc \\ 6bbcc \\ 4bccc \\ cccc \end{cases}
$$

19

$aaaa + daaa - ggga = bbbb.$

$$A \begin{cases} -ggg \\ d \\ bbb \end{cases} \quad B \begin{cases} -ggg \\ d \\ 3\,db \\ 3\,dbb \\ 4\,bbb \\ 6\,bb \\ 4\,b \end{cases}$$

$$Ab \begin{cases} -gggb \\ dbbb \\ bbbb \end{cases} \quad Bc \begin{cases} -gggc \\ 3\,dbcc \\ 3\,dbbc \\ 4\,bbbc \\ 6\,bbcc \\ 4\,bccc \\ cccc \end{cases}$$

20

$-aaaa - daaa + ggga = bbbb.$

$$A \begin{cases} ggg \\ -d \\ -bbb \end{cases} \quad B \begin{cases} ggg \\ -d \\ -3\,db \\ -3\,dbb \\ -4\,bbb \\ -6\,bb \\ -4\,b \end{cases}$$

$$Ab \begin{cases} gggb \\ -dbbb \\ -bbbb \end{cases} \quad Bc \begin{cases} gggc \\ -dccc \\ -3\,dbcc \\ -3\,dbbc \\ -4\,bbbc \\ -6\,bbcc \\ -4\,bccc \end{cases}$$

21

$aaaa - daaa + ggga = bbbb.$

$$A \begin{cases} ggg \\ -d \\ bbb \end{cases} \quad B \begin{cases} ggg \\ -d \\ -3\,db \\ -3\,dbb \\ 4\,bbb \\ 6\,bb \\ 4\,b \end{cases}$$

$$Ab \begin{cases} gggb \\ -dbbb \\ bbbb \end{cases} \quad Bc \begin{cases} ggggc \\ -dccc \\ -3\,dbcc \\ -3\,dbbc \\ 4\,bbbc \\ 6\,bbcc \\ 4\,bccc \\ cccc \end{cases}$$

22

$-aaaa + daaa - ggga = bbbb.$

$$A \begin{cases} -ggg \\ a \\ -bbb \end{cases} \qquad B \begin{cases} -ggg \\ d \\ 3\,db \\ 3\,dbb \\ -4\,bbb \\ -6\,bb \\ -4\,b \end{cases}$$

$$Ab \begin{cases} -gggb \\ dbbb \\ -bbbb \end{cases}$$

$$Bc \begin{cases} -gggc \\ dccc \\ 3\,dbcc \\ 3\,dbbc \\ -4\,bbbc \\ -6\,bbcc \\ -4\,bccc \\ -cccc \end{cases}$$

23

$aaaa - daaa - ggga = bbbb.$

$$A \begin{cases} -ggg \\ -d \\ -bbb \end{cases} \qquad B \begin{cases} -ggg \\ -d \\ -3\,db \\ -3\,dbb \\ 4\,bbb \\ 6\,bb \\ 4\,b \end{cases}$$

$$Ab \begin{cases} -gggb \\ -dbbb \\ bbbb \end{cases}$$

$$Bc \begin{cases} -gggc \\ -dccc \\ -3\,dbcc \\ -3\,dbbc \\ 4\,bbbc \\ 6\,bbcc \\ 4\,bccc \\ cccc \end{cases}$$

24

$-aaaa + daaa + gggz = bbbb.$

$$A \begin{cases} ggg \\ d \\ -bbb \end{cases} \quad B \begin{cases} ggg \\ d \\ 3\,db \\ 3\,dbb \\ -4\,bbb \\ -6\,bb \\ -4\,b \end{cases}$$

$$Ab \begin{cases} gggb \\ dbbb \\ -bbbb \end{cases}$$

$$Bc \begin{cases} gggc \\ dccc \\ 3\,dbcc \\ 3\,dbbbc \\ -4\,bbbc \\ -6\,bbcc \\ -4\,bccc \\ -cccc \end{cases}$$

25

175

$aaaa + daaa + ggga = bbbb.$

$$A \begin{cases} ff \\ d \\ bbb \end{cases} \quad B \begin{cases} ff \\ 2\,ffb \\ d \\ 3\,db \\ 3\,dbb \\ 4\,bbb \\ 6\,bb \\ 4\,b \end{cases}$$

$$Ab \begin{cases} ffbb \\ dbbb \\ bbbb \end{cases}$$

$$Bc \begin{cases} ffcc \\ 2\,ffbc \\ dccc \\ 3\,dbcc \\ 3\,dbbc \\ 4\,bbbc \\ 6\,bbcc \\ 4\,bccc \\ cccc \end{cases}$$

26

$$aaaa + daaa - ffaa = bbbb.$$

$$
A \begin{cases} -ff \\ d \\ bbb \end{cases}
\qquad
B \begin{cases} -ff \\ -2\,ffb \\ d \\ 3\,db \\ 3\,dbb \\ 4\,bbb \\ 6\,bb \\ 4\,b \end{cases}
$$

$$
Ab \begin{cases} -ffbb \\ dbbb \\ bbbb \end{cases}
$$

$$
Bc \begin{cases} -ffcc \\ -2\,ffbc \\ dccc \\ 3\,dbcc \\ 3\,dbbc \\ 4\,bbbc \\ 6\,bbcc \\ 4\,bccc \\ cccc \end{cases}
$$

27

$$-aaaa - daaa + ffaa = bbbb.$$

$$
A \begin{cases} ff \\ -d \\ -bbb \end{cases}
\qquad
B \begin{cases} ff \\ 2\,ffb \\ -d \\ -3\,db \\ -3\,dbb \\ -4\,bbb \\ -6\,bb \\ -4\,b \end{cases}
$$

$$
Ab \begin{cases} ffbb \\ -dbb \\ -bbbb \end{cases}
$$

$$
Bc \begin{cases} ffcc \\ 2\,ffbc \\ -dccc \\ -3\,dbcc \\ -3\,dbbc \\ -4\,bbbc \\ -6\,bbcc \\ -4\,bccc \\ -cccc \end{cases}
$$

28

$aaaa - daaa + ffaa = bbbb.$

$$A \begin{cases} ff \\ -d \\ bbb \end{cases}$$

$$Ab \begin{cases} ffbb \\ -dbbb \\ bbbb \end{cases}$$

$$B \begin{cases} ff \\ 2\,ffb \\ -d \\ -3\,db \\ -3\,dbb \\ 4\,bbb \\ 6\,bb \\ 4\,b \end{cases}$$

$$Bc \begin{cases} ffcc \\ 2\,ffbc \\ -dccc \\ -3\,dbcc \\ -3\,dbbc \\ 4\,bbbc \\ 6\,bbcc \\ 4\,bccc \\ cccc \end{cases}$$

176

29

$-aaaa + daaa - ffaa = bbbb.$

$$A \begin{cases} -ff \\ d \\ -bbb \end{cases}$$

$$Ab \begin{cases} -ffbb \\ dbbb \\ -bbbb \end{cases}$$

$$B \begin{cases} -ff \\ -2\,ffb \\ d \\ 3\,db \\ 3\,dbb \\ -4\,bbb \\ -6\,bb \\ -4\,b \end{cases}$$

$$Bc \begin{cases} -ffcc \\ -2\,ffbc \\ dccc \\ 3\,dbcc \\ 3\,dbbc \\ -4\,bbbc \\ -6\,bbcc \\ -4\,bccc \\ cccc \end{cases}$$

30

$-aaaa + daaa + ffaa = bbbb.$

$$A \begin{cases} ff \\ d \\ -bbb \end{cases}$$

$$Ab \begin{cases} ffbb \\ dbbb \\ -bbbb \end{cases}$$

$$B \begin{cases} ff \\ 2\,ffb \\ d \\ 3\,db \\ 3\,dbb \\ -4\,bbb \\ -6\,bb \\ -4\,b \end{cases}$$

$$Bc \begin{cases} ffcc \\ 2\,ffbc \\ dccc \\ 3\,dbcc \\ 3\,dbbc \\ -4\,bbbc \\ -6\,bbcc \\ -4\,bccc \\ -cccc \end{cases}$$

31

$aaaa - daaa + ffaa = bbbb.$

$$A \begin{cases} -ff \\ -d \\ bbb \end{cases}$$

$$Ab \begin{cases} -ffbb \\ -dbbb \\ bbbb \end{cases}$$

$$B \begin{cases} -ff \\ -2\,ffb \\ -d \\ -3\,db \\ -3\,dbb \\ 4\,bbb \\ 6\,bb \\ 4\,b \end{cases}$$

$$Bc \begin{cases} -ffcc \\ -2\,ffb \\ -dccc \\ -3\,dbcc \\ -3\,dbbc \\ 4\,bbbc \\ 6\,bbcc \\ 4\,bccc \\ cccc \end{cases}$$

32

$aaaa + daaa + ffaa + ggga = bbbb.$

$$
A \begin{cases} gggg \\ ff \\ d \\ bbb \end{cases}
$$

$$
Ab \begin{cases} gggb \\ ffbb \\ dbbb \\ bbbb \end{cases}
$$

$$
B \begin{cases} ggg \\ ff \\ 2\,ffb \\ d \\ 3\,db \\ 3\,dbb \\ 4\,bbb \\ 6\,bb \\ 4\,b \end{cases}
$$

$$
Bc \begin{cases} gggc \\ ffcc \\ 2\,ffbc \\ dccc \\ 2\,dbcc \\ 3\,dbbc \\ 4\,bbbc \\ 6\,bbcc \\ 4\,bccc \\ cccc \end{cases}
$$

---

33   **177**

$aaaa + daaa + ffaa - ggga = bbbb.$

$$
A \begin{cases} -ggg \\ ff \\ d \\ bbb \end{cases}
$$

$$
Ab \begin{cases} -gggb \\ ffbb \\ dbbb \\ bbbb \end{cases}
$$

$$
B \begin{cases} -ggg \\ ff \\ 2\,ffb \\ d \\ 3\,db \\ 3\,dbb \\ 4\,bbb \\ 6\,bb \\ 4\,b \end{cases}
$$

$$
Bc \begin{cases} -gggc \\ ffcc \\ 2\,ffbc \\ dccc \\ 3\,dbcc \\ 3\,dbbc \\ 4\,bbbc \\ 6\,bbcc \\ 4\,bccc \\ cccc \end{cases}
$$

34

$$-aaaa - daaa - ffaa + ggga = bbbb.$$

$$A \begin{cases} ggg \\ -ff \\ -d \\ -bbb \end{cases} \quad Ab \begin{cases} gggb \\ -ffbb \\ -dbbb \\ -bbbb \end{cases} \quad B \begin{cases} ggg \\ -ff \\ -2\,ffb \\ -d \\ -3\,db \\ -3\,dbb \\ -4\,bbb \\ -6\,bb \\ -4\,b \end{cases} \quad Bc \begin{cases} gggc \\ -ffcc \\ -2\,ffbc \\ -dccc \\ -3\,dbcc \\ -3\,dbbc \\ -4\,bbbc \\ -6\,bbcc \\ -4\,bccc \\ -cccc \end{cases}$$

35

$$aaaa + daaa - ffaa + ggga = bbbb.$$

$$A \begin{cases} ggg \\ -ff \\ d \\ bbb \end{cases} \quad Ab \begin{cases} gggb \\ -ffbb \\ dbbb \\ bbbb \end{cases} \quad B \begin{cases} ggg \\ -ff \\ -2\,ffb \\ d \\ 3\,db \\ 4\,bbb \\ 6\,bb \\ 4\,b \end{cases} \quad Bc \begin{cases} gggc \\ -ffcc \\ -2\,ffbc \\ dccc \\ 3\,dbcc \\ 3\,dbbc \\ 4\,bbbc \\ 6\,bbcc \\ 4\,bccc \\ cccc \end{cases}$$

36

$$-aaaa - daaa + ffaa - ggga = bbbb.$$

$$A \begin{cases} -ggg \\ ff \\ -d \\ -bbb \end{cases}$$

$$Ab \begin{cases} -gggb \\ ffbb \\ -dbbb \\ -bbbb \end{cases}$$

$$B \begin{cases} -ggg \\ ff \\ 2\,ffb \\ -d \\ -3\,db \\ -3\,dbb \\ -4\,bbb \\ -6\,bb \\ -4\,b \end{cases}$$

$$Bc \begin{cases} gggc \\ ffcc \\ 2\,ffbc \\ -dccc \\ -3\,dbcc \\ -3\,dbbc \\ -4\,bbbc \\ -6\,bbcc \\ -4\,bccc \\ -cccc \end{cases}$$

---

37   **178**

$$aaaa - daaa + ffaa + ggga = bbbb.$$

$$A \begin{cases} ggg \\ ff \\ -d \\ bbb \end{cases}$$

$$Ab \begin{cases} gggb \\ ffbb \\ -dbbb \\ bbbb \end{cases}$$

$$B \begin{cases} ggg \\ ff \\ 2\,ffb \\ -d \\ -3\,db \\ -3\,dbb \\ 4\,bbb \\ 6\,bb \\ 4\,b \end{cases}$$

$$Bc \begin{cases} gggc \\ ffcc \\ 2\,ffbc \\ -dccc \\ -3\,dbcc \\ -3\,dbbc \\ 4\,bbbc \\ 6\,bbcc \\ 4\,bccc \\ cccc \end{cases}$$

38

$$-aaaa + daaa - ffaa - ggga = bbbb.$$

$$
A \begin{cases} -ggg \\ -ff \\ d \\ -bbb \end{cases}
\quad
Ab \begin{cases} -gggb \\ -ffbb \\ dbbb \\ -bbbb \end{cases}
\quad
B \begin{cases} -ggg \\ -ff \\ -2\,ffb \\ d \\ 3\,db \\ 3\,dbb \\ -4\,bbb \\ -6\,bb \\ -4\,b \end{cases}
\quad
Bc \begin{cases} -gggc \\ -ffcc \\ -2\,ffbc \\ dccc \\ 3\,dbcc \\ 3\,dbbc \\ -4\,bbbc \\ -6\,bbcc \\ -4\,bccc \\ -cccc \end{cases}
$$

39

$$aaaa + daaa - ffaa - ggga = bbbb.$$

$$
A \begin{cases} -ggg \\ -ff \\ d \\ bbb \end{cases}
\quad
Ab \begin{cases} -ggb \\ -ffbb \\ dbbb \\ bbbb \end{cases}
\quad
B \begin{cases} -ggg \\ -ff \\ -2\,ffb \\ d \\ 3\,db \\ 3\,abb \\ 4\,bbb \\ 6\,bb \\ 4\,b \end{cases}
\quad
Bc \begin{cases} -gggc \\ -ffcc \\ -2\,ffbc \\ dccc \\ 3\,dbcc \\ 3\,dbbc \\ 4\,bbbc \\ 6\,bbcc \\ 4\,bccc \\ cccc \end{cases}
$$

40

$$-aaaa - daaa + ffaa + ggga = bbbb.$$

A $\left\{ \begin{array}{c} ggg \\ ff \\ -d \\ -bbb \end{array} \right.$

Ab $\left\{ \begin{array}{c} ggb \\ ffbb \\ -dbbb \\ -bbbb \end{array} \right.$

B $\left\{ \begin{array}{c} ggg \\ ff \\ 2ffb \\ -d \\ -3db \\ -3dbb \\ -4bbb \\ -6bb \\ -4b \end{array} \right.$

Bc $\left\{ \begin{array}{c} gggc \\ ffcc \\ 2ffbc \\ -dccc \\ -3dbcc \\ -3dbbc \\ -4bbbc \\ -6bbcc \\ -4bccc \\ -cccc \end{array} \right.$

---

41   **41**

$$aaaa - daaa + ffaa - ggga = bbbb.$$

A $\left\{ \begin{array}{c} -ggg \\ ff \\ -d \\ bbb \end{array} \right.$

Ab $\left\{ \begin{array}{c} -gggb \\ ffbb \\ -dbbb \\ bbbb \end{array} \right.$

B $\left\{ \begin{array}{c} -ggg \\ ff \\ 2ffb \\ -d \\ -3db \\ -3dbb \\ 4bbb \\ 6bb \\ 4b \end{array} \right.$

Bc $\left\{ \begin{array}{c} -gggc \\ ffcc \\ 2ffbc \\ -dccc \\ -3dbcc \\ -3dbbc \\ 4bbbc \\ 6bbcc \\ 4bccc \\ cccc \end{array} \right.$

42

$$-aaaa + daaa - ffaa + ggga = bbbb.$$

$$A \begin{cases} ggg \\ -ff \\ d \\ -bbb \end{cases}$$

$$Ab \begin{cases} gggb \\ -ffbb \\ dbbb \\ -bbbb \end{cases}$$

$$B \begin{cases} ggg \\ -ff \\ -2ffb \\ d \\ 3db \\ 3dbb \\ -4bbb \\ -6bb \\ -4b \end{cases}$$

$$Bc \begin{cases} gggc \\ -ffcc \\ -2ffbc \\ dccc \\ 3dbcc \\ 3dbbc \\ -4bbbc \\ -6bbcc \\ -4bccc \\ -cccc \end{cases}$$

43

$$aaaa - daaa - ffaa + ggga = bbbb.$$

$$A \begin{cases} ggg \\ -ff \\ -d \\ bbb \end{cases}$$

$$Ab \begin{cases} gggb \\ -ffbb \\ -dbbb \\ bbbb \end{cases}$$

$$B \begin{cases} gggg \\ -ff \\ -2ffb \\ -d \\ -3db \\ -3dbb \\ 4bbb \\ 6bb \\ 4b \end{cases}$$

$$Bc \begin{cases} gggc \\ -ffcc \\ -2ffb \\ -dccc \\ -3dbcc \\ -3dbbc \\ 4bbbc \\ 6bbcc \\ 4bccc \\ cccc \end{cases}$$

44

$$-aaaa + daaa + ffaa - ggga = bbbb.$$

$$A \begin{cases} -ggg \\ ff \\ d \\ -bbb \end{cases} \qquad B \begin{cases} -ggg \\ ff \\ 2ffb \\ d \\ 3db \\ 3dbb \\ -4bbb \\ -6bb \\ -4b \end{cases}$$

$$Ab \begin{cases} -gggb \\ ffbb \\ dbbb \\ -bbbb \end{cases}$$

$$Bc \begin{cases} -gggc \\ ffcc \\ 2ffbc \\ dccc \\ 3dbcc \\ 3dbbc \\ -4bbbc \\ -6bbcc \\ -4bccc \\ cccc \end{cases}$$

---

45   **180**

$$aaaa - daaa - ffaa - ggga = bbbb.$$

$$A \begin{cases} -ggg \\ -ff \\ -d \\ bbb \end{cases} \qquad B \begin{cases} -ggg \\ -ff \\ -2ffb \\ -d \\ -3db \\ -3dbb \\ 4bbb \\ 6bb \\ 4b \end{cases}$$

$$Ab \begin{cases} -gggb \\ -ffbb \\ -dbbb \\ bbbb \end{cases}$$

$$Bc \begin{cases} -gggc \\ -ffcc \\ -2ffbc \\ -dccc \\ -3dbcc \\ -3dbbc \\ 4bbbc \\ 6bbcc \\ 4bccc \\ cccc \end{cases}$$

$$46$$

$$-aaaa + daaa + ffaa + ggga = bbbb.$$

$$
A\begin{cases} ggg \\ ff \\ d \\ -bbb \end{cases}
\qquad
B\begin{cases} ggg \\ ff \\ 2\,ffb \\ d \\ 3\,db \\ 3\,dbb \\ -4\,bbb \\ -6\,bb \\ -4\,b \end{cases}
$$

$$
Ab\begin{cases} gggb \\ ffbb \\ dbbb \\ -bbbb \end{cases}
$$

$$
Bc\begin{cases} gggc \\ ffcc \\ 2\,ffbc \\ dccc \\ 3\,dbcc \\ 3\,dbbc \\ -4\,bbbc \\ -6\,bbcc \\ -4\,bccc \\ -cccc \end{cases}
$$

180 And that brings to a close the complete description of equations in the three proposed classes, together with their canons of direction. As a general corollary of the earlier problems, and as a very appropriate appendix for the numerical exegesis, this synopsis also finishes off the second, exegetic part of this treatise.

To students of mathematics

It was only after much consideration that, out of all of Thomas Harriot's mathematical writings, this analytical work was published first. His other works (which also stand out for their abundant novelty of invention) were composed in quite the same logistic style as this treatise, which consists entirely of examples of every variety of specious logistic. Such a style was, until now, completely unknown. And that is reason why this treatise — quite apart from its own inestimable utility — might serve as a kind of necessary preparation or introduction, a forerunner, as it were, to the rest of Harriot's writings (the publication of which is now under serious consideration). We thought it would be worthwhile, in this brief note, to forewarn students of mathematics that this present work will be useful for approaching those other writings.

Commentary

Notes on Preface to Analysts

1. That is, the methods of Euclid's *Elements*, using only compass and straightedge.
2. Synonymous to compass and straightedge constructions and equivalent, in algebraic terms to the solution of a quadratic equation.
3. See Knorr, 1986, 341–8 for more on this subject.
4. Reference is made to Viète's works: *De aequationum geomtriae* (first published 1593); *De aequationum recognitione et emendatione* (1615); and *Ad angularium sectionum analyticen theoremata* (1615).

Notes on the Definitions

1. This Definition may be compared with that in Viète's *In Artem Analyticem Isagoge*, 1600, p. 5 Chapter IIII.

"Numerical logistic is [a logistic] that employs numbers, symbolic logistic is one that employs symbols or signs for things as, say, the letters of the alphabet". (Witmer, 1983, p. 17).[1]

The rules governing operations between species are to be found in the same chapter of Viète's work. They begin with the following:

RULE I: To Add One Magnitude to Another

Let there be two magnitudes, A and B. One is to be added to the other.

Since one magnitude is to be added to another, and homogeneous and heterogeneous terms do not affect each other, the two magnitudes proposed are homogeneous. (Greater or less do not constitute differences in kind.) Therefore, they will

---

[1] De praeceptis Logistices speciosae.
Logistice numerosa est quae per numeros, Speciosa quae per species seu rerum formas exhibetur, utpote per Alphabetica elementa.
Logistices speciosae canonica praecepta sunt quatuor, ut numerosae.

be properly added by the signs of conjunction or addition and their sum will be A plus B, if they are simple lengths or breadths. But if they are higher up in the series set out above or if, by their nature they correspond to higher terms, they should be properly designated as, say A^2 plus B^P, or A^3 plus B^S, and so forth for the rest.

Analysts customarily indicate a positive affection by the symbol $+$.

RULE II: To Subtract One Magnitude from Another

Let there be two magnitudes, A and B, the former the greater, the latter the less. The smaller is to be subtracted from the greater.

Since one magnitude is to be subtracted from another and homogeneous and heterogeneous magnitudes do not affect each other, the two magnitudes given are homogeneous. (Greater or less do not constitute differences in kind.) Therefore, the subtraction of the smaller from the larger is properly made by the sign of disjunction or subtraction, and the disjoint terms will be A minus B if they are only simple lengths or breadths. But if they are higher up in the series set out above or if, by their nature, they correspond to higher terms, they should be properly designated as, say, A^2 minus B^P, or A^3 minus B^S, and so forth for the rest.[2]

The corresponding rules may be seen below in Section 1 (The Forms of the Four Operations of Specious Arithmetic) and in the Harriot manuscripts (Add. MS 6782, ff. 322–5), which almost replicate Section 1 of the *Praxis*. Another version exists in the pages of the Torporley papers in Lambeth Palace library. It is clear that Harriot and Warner have "algebraicized" Viète. Viète's "magnitudes," "lengths," and "breadths" have no place in Harriot's algebra.

This Definition is the first of several (including Synthesis, Analysis, Zetetic, Exegetic, and Poristic) that form the subject matter of the *Isagoge*, and these five

---

[2] PRAECEPTVM I.
Magnitudinem magnitudini addere.
Sunto duae magnitudines A & B. Oportet alteram alteri addere.

Quoniam igitur magnitudo magnitudini addenda est, homogeneae autem heterogeneas non adficiunt, sunt quae proponuntur addendae duae magnitudines homogeneae. Plus autem vel minus non constituunt genera diversa. Quare nota copulae, seu adiunctionis commode addentur & adgregatae erunt A plus B, siquidem sint simplices longitudines latitudinesve. Sed si adscendant per expositam scalam, vel adscendentibus genere communicant, sua quae congruit designabuntur denominatione, veluti dicetur A quadratum plus B plano, vel A cubus plus B solido, & similiter in reliquis.

Solent autem Analystae symbolo - - - - adfectionem adiunctionis indicare.
PRAECEPTUM II.
Magnitudinem magnitudini subducere.
Sunto duae magnitudines A & B, illa major, haec minor. Oportet minorem a maiore subducere.

Quoniam igitur magnitudo magnitudini subducenda est, homogeneae autem magnitudines heterogeneas non adficiunt, sunt quae proponuntur duae magnitudines homogeneae. Plus autem vel minus non constituunt genera diuersa. Quare nota disiunctionis seu multae commode minoris a maiore fiet subductio & disiunctae erunt A minus B, siquidem sint simplices longitudines latitudinesve.

Sed si adscendant per expositam scalam vel adscendentibus genere communicant, sua quae congruit designabuntur denominatione, veluti dicetur A quadratum minus B plano, vel A cubus minus B.

concepts are also to be found in the manuscript papers (Add. MS 6784, ff. 19–28 in the British Library).

These latter papers present geometric problems solved by predominantly algebraic methods using mostly algebraic notation, although not exclusively. ff. 19–23 start from a pair of intersecting straight lines and (rather like Descartes' investigation of Pappus' Locus Problem) look for what happens when the product of two line-segments is constant. ff. 24–28 investigate the rational section of a straight line by two intersecting straight lines. ("De sectio rationis")

Some geometrical notation is used throughout but the "algebraicization" of the problems, including the formation of equations suggests strongly that Harriot was feeling his way (at the very least) towards analytical geometry.

Fortunately, we have the manuscript pages and it is clear that the concepts as used by Harriot and defined by Warner are rooted in the *Isagoge* although both Harriot and Warner, in their practical usage, transcend the usage of Viète and give these terms an exclusively algebraic meaning.

2. Synthesis is referred to as "the best method of proof" for propositions of any sort and is to be found below in Section 2 in which the synthetic method is used to generate Canonical Equations. The corresponding Viète definition of synthesis and other terms is found in Chapter I of the *Isagoge*.

"There is a certain way of searching for the truth in mathematics that Plato is said first to have discovered. Theon called it analysis, which he defined as assuming that which is sought as if it were admitted [and working] through the consequences [of that assumption] to what is admittedly true, as opposed to synthesis, which is assuming what is [already] admitted [and working] through the consequences [of that assumption] to arrive at and to understand that which is sought.

Although the ancients propounded only [two kinds of] analysis, zetetics and poristics, to which the definition of Theon best applies, I have added a third, which may be called rhetics or exegetics. It is properly zetetics by which one sets up an equation or proportion between a term that is to be found and the given terms, poristics by which the truth of a stated theorem is tested by means of an equation or proportion and exegetics by which the value of the unknown term in a given equation or proportion is determined. Therefore the whole analytic art, assuming this threefold function for itself, may be called the science of correct discovery in mathematics".[3]

Add. MS 6784, ff. 27–8 shows Harriot's use of synthesis.

---

[3] Est veritatis inquirendae via quaedam in Mathematicis quam Plato primus invenisse dicitur, a Theone nominata Analysis, & ab eodem definita, Adsumptio quaesiti tanquam concessi per consequentia ad verum concessum. Vt contra Synthesis, Adsumptio concessi per consequentia ad quaesiti finem & comprehensionem. Et quanquam veteres duplicem tantum proposuerunt Analysim ζητητικὴν καὶ ποριστικὴν ad quas definitio Theonis maxime pertinet, constitui tamen etiam tertiam speciem, quae dicatur ῥητικὴ ἢ ἐξηγητική, consentaneum est, ut sit Zetetice qua invenitur aequalitas proportiove magnitudinis de qua quaeritur cum iis quae data sunt. Poristice, qua de aequalitate vel proportione ordinate Theorematis veritas examinatur. Exegetice, qua ex ordinata aequalitate vel proportione ipsa de qua quaeritur exhibetur magnitudo. Atque adeo tota ars Analytice triplex illud sibi vendicans officium definiatur, doctrina bene inveniendi in Mathematicis.

3. This definition of analysis may be compared with that in the *Isagoge*, and it will be seen that the references in the *Praxis* are to quantity, power and coefficients without reference to magnitude. Viète's algebra referred to magnitude and, thus, had a geometric aspect. See also Add. MS 6784, f. 27 in which Harriot refers the term to a purely algebraic problem.

4. Viète's definition of analysis had distinguished three types: Zetetic (setting up equations), Exegetic (solution of equations) and Poristic (investigating the truth of the result). Definition 6 (of Zetetic) again demonstrates a move to pure algebra in its reference to quantity rather than magnitude (as in Viète). See Add. MS 6784, f. 22 in which Harriot's work, headed Zetetic, is exclusively algebraic. See also Add. MS 6784, ff. 19, 24, 25 for further Harriot references to Zetetic.

Finally *Isagoge*, Chapter V should be compared with the second part of Section 1 (below), both of which give the Laws of Zetetics, in order to appreciate the difference made by the new notation.

5. The geometrical aspect in Viète's work has given way to a purely algebraic treatment. Compare *Isagoge*, Problem II, *Praxis*, Problem VII and Add. MS 6782, f. 394 that provide solutions to the same equation with numerical coefficients. (See Table below, pp. 262).

6. For Viète's definition of Poristic, see above, p. 211. In Chapter VI of the *Isagoge*, Viète refers to Poristic in its geometric mode by giving examples in Theon, Apollonius, and Archimedes. The distinguishing characteristic of Poristic given in Definition 8 is that it leads to an identity or a given quantity and not an equality, as in Zetetic. For Harriot's own work see Add. MS 6784, ff. 20, 21, 25. In Harriot's and Warner's hands, Poristic is algebraic. Section 4 of the *Praxis* investigates the truth of previous propositions and each investigation ends in an identity. Thus, Section 4 of the *Praxis* is Poristic.

Notes on Section One

1. The first part of this Section corresponds to Chapter IIII of Viète's *Isagoge*, "On the Rules of Symbolic Logistic."

The operational rules laid out below, although incomplete, combine with Harriot's symbolic notation to enable the 'algebraic' logic to emerge in a similar way to that in which geometric rigour had operated for so long.

Viète's use of vowels and consonants for unknowns and knowns respectively, but with lower case letters, is copied throughout. The signs + and − and = are used although Harriot's usual sign for multiplication (when an algebraic manipulation is intended) does not appear in the *Praxis* until p. 10. The division of a fraction by a whole number or other fraction is denoted on p. 10 by two parallel lines rather like equality, but somewhat smaller and easily distinguishable from it.

Zero is given numerical status first on p. 8 (of the *Praxis*) as the result of subtraction and, later in the work, will be followed by "equating to zero" in the arrangement of equations. The Harriot manuscripts take this even further and there are countless examples of homogeneity being maintained even to the extent of writing four zeros on the right-hand side of an equation of the eighth degree (e.g., Add. MS 6783, f. 171). (This is on a page listing three canonical equations with their "originals.") It is all the more intriguing that on Add. MS 6782, ff. 91–93 there are arithmetical calculations in which zero is missing, e.g.,

$$
\begin{array}{r}
17 \\
\underline{31} \\
17 \\
\underline{51} \\
527
\end{array}
$$

This appears on a page of miscellaneous "rough" calculations.

On f. 93, dotted guidelines are used instead. Letters stand for positive numbers throughout the *Praxis*, although the editor(s) slip from this consistency on a few occasions. Pp. 95 and 97 will give negative roots and pp. 7 and 8 of this Section show examples of addition and subtraction of negative numbers.

The lack of exponent is already clear in this Section and, although it means that algebraic laws and results might not be expressed most economically, the resulting necessity for counting, although tedious, does not detract from the purely quantitative nature of the notation. Nowhere in the Harriot manuscripts is this practice departed from and, as Wallis suggested (*Algebra* 1685, p. 126) the "ambiguity inherent in previous writers' work using cossic notation, was thereby removed". Such ambiguity surely referred to knowing whether an additive or a multiplicative system of powers was being used.

Throughout the book, all equations are in homogenous form (the first case is on p. 11 of this Section). Despite this, the ancient Greek dimensional limitation is not adhered to. Algebraic quantities have no geometrical associations in the *Praxis*. It will be seen below that equations are treated differently, according to "type", i.e., according to the signs of the coefficients and the powers involved.

Amongst the unpublished MS papers are four pages (Add. MS 6784, ff. 322–325) that are very similar to Section 1, the minor alterations in the *Praxis* having been made, in all likelihood, for purposes of clarity. Torporley's *Summary* has a version of Section 1. (Sion College MS Arc. L.40.2/L.40, f. 35–54v.)

Section 1 provides the first opportunity to compare the contents of the *Praxis* with those of the MS and Torporley's copy. Section 1, giving some basic algebraic rules, is headed "Logistices Speciosae quatuor operationum formae practicae exemplis declarantur" (The Forms of the Four Operations of Specious Arithmetic Illustrated by Example). It is difficult to judge why Warner changed the original heading given in both the MS and Torporley copy: "Operationes Logisticae in Notis".

The first of the manuscript pages f. 322 shows an arrangement of addition and subtraction examples different from the book and halfway down the page (in subtracting $c - d$ from $a + b$) a comma is used as a sort of bracket. F. 323 shows more divergences from the *Praxis*. In the first line, examples are interchanged and in the top right-hand corner is a case not appearing in Section 1. The next line again uses a comma as a sort of bracket (not in the *Praxis*) and the numerical example in the next line does not appear in Section 1. Several further examples on f. 323 are discarded in the *Praxis* but an extra example on p. 9 of the latter (*bcdc* divided by *bdc*) is not to be found in the papers. F. 323 uses, at the bottom of the page, the uniquely characteristic sign used by Harriot throughout the papers for equality, never making an appearance in the *Praxis* (except for an echo on pp. 63, 4).

Add. MS 6784 f. 324 corresponds to *Praxis*, p. 10, most of the changes being relatively minor. A major difference, however, lies in the absence on f. 324 of any reference to equality or inequality signs. An extra example of reduction of fractions to their lowest terms is added in Section 1 ($bcda/ca = bd$) and throughout the page d replaces z in the manuscript. In the bottom left-hand corner the example is replicated in the *Praxis* but with c in place of z and g in place of c.

Three final cases on this page (bottom right) do not appear in Section 1. F. 325 presents similar discrepancies, some change of letters and some minor alterations in the first and last examples. We can only speculate as to whether Warner had seen these actual pages or worked from others. But since the book very probably had a teaching purpose, clarity and simplicity may have been in the forefront of his mind.

The Torporley copy differs more from the Harriot manuscripts than the *Praxis* does. Examples of the four rules are given but they are not the same examples as in the Harriot papers, f. 35r has:

$$b + c - d$$
$$b - c + d$$
$$\overline{bb + bc - bd}$$
$$-bc - cc + cd$$
$$\underline{\qquad +bd + cd - dd}$$
$$bb - cc + 2, cd - dd$$

Next to this is written:

$$8 - 2$$
$$\underline{8 - 2}$$
$$64 - 16$$
$$\underline{-16 + 4}$$
$$68 - 32 \qquad = 36$$

35v has work towards the top of the page not (yet) found in the Harriot manuscripts.

The material about Antithesis, Hypobibasmus and Parabolismus originally taken from Viète is the same here as in manuscripts and in the *Praxis*.

Torporley's version of Section 1 is followed by a section headed *De Radicalibus* that does not appear at all in the *Praxis* but has pages correspondingly headed in Harriot's papers.

Warner decided to omit *De Radicalibus* and it is more difficult in this case to find convincing reasons for his having done so than to find reasons for altering and cutting down the material for Section 1.

2. See Definition 1 for Warner's idea of the meaning of this term.

3. Throughout the *Praxis* there appears to be a trend towards generalization through lists of examples, which are, all too often, incomplete. One case of such "incomplete generalization" occurs in Section 1. However, Warner's use of "exemplificatae" implies a knowledge that not all cases may be given and that those displayed will be sufficient to generalize from.

4. Viète denotes "removal" by – and verbally refers back to "defect" ($\lambda\epsilon\tilde{\iota}\psi\iota\varsigma$) in Diophantus. (*Isagoge*, Chapter IV, Rule II.)

5. It is at this point in the original that a dot is used between numerical and literal parts of a term. The use is not consistent and has had to be omitted in the present edition for practical reasons.

6. Note the absence here of Harriot's normal sign for multiplication (⌐), always used for calculation in the papers and used, too, from p. 10 onwards in the *Praxis*. However, on f. 324 of Add. MS 6782, he uses (⌐).

7. The word "application" is a carry-over from the Greek geometrical use of the term. T. L. Heath's edition of the *Elements* (Vol. 1, p. 343) quotes Proclus' (following Eudemus) explanation of the method, which enables "division" to be done geometrically.

"For, when you have a straight line set out and lay the given area exactly alongside the whole of the straight line, then they say that you *apply* ($\pi\alpha\rho\alpha\beta\dot\alpha\lambda\lambda\epsilon\iota\nu$) the said area; when however you make the length of the area greater than the straight line itself, it is said to *exceed* ($\dot{Y}\pi\epsilon\rho\beta\dot\alpha\lambda\lambda\epsilon\iota\nu$), and when you make it less, in which case after the area has been drawn, there is some part of the straight line extending beyond it, it is said to *fall short* ($\dot{E}\lambda\lambda\epsilon\dot\iota\pi\epsilon\iota\nu$). Euclid too, in the sixth book, speaks in this way both of *exceeding* and *falling-short*; but in this place he needed the *application* simply, as he sought to apply to a given straight line an area equal to a given triangle in order that we might have in our power, not only the *construction* (σύστασις) of a parallelogram equal to a given triangle, but also the *application* of it to a finite straight line. For example, given a triangle with an area of 12 feet, and a straight line set out the length of which is 4 feet, we apply to the straight line the area equal to the triangle if we take the whole length of 4 feet and find how many feet the breadth must be in order that the parallelogram may be equal to the triangle. In the particular case, if we find breadth of 3 feet and multiply the length into the breadth, supposing that the angle set out is a right angle, we shall have the area. Such then is the *application* handed down from early times by the Pythagoreans".

8. The three words are not used at all below.

9. The inequality signs shown here are never used again in a totally identical form. In Sections 5 and 6 the signs used are capital V on its side (see pp. 78–88). The papers throughout use [⊲] and [⊳] without exception. (See especially Add. MS 6783, ff. 106, 184 and 185. These pages are devoted to the inequalities which appear in a slightly different form in Section 5 of the *Praxis*, pp. 78–86). Also, note Add. MS 6782, ff. 187r, 189r, 190r and 192r for an idiosyncratic sign Ꞓ whose meaning is not clear. These pages are covered with algebraic manipulations whose purpose is unclear and which may be part of the "waste" referred to by Harriot in his Will. (It is also noteworthy that on Add. MS 6783, f. 116 Harriot uses "$\pm$" in a totally modern manner in front of the terms of an equation, remarkable for its time. This page, headed *On Solving Equations by Reduction*, solves a pair of conjugate biquadratic equations by a method which follows that of Viète. Ⅱ consistently denotes equality in the MS papers. The bars are usually short but sometimes long (Add. MS 6782, f. 262, a page on which Harriot is using algebra to solve the geometrical problem of dividing a line in extreme and mean ratio). The bars are usually horizontal but sometimes vertical (e.g., Add. MS 6782, f. 14r, f. 15r, which are pages covered in numerical calculations, probably "rough" work) and there are pages that have some algebra thrown in. Horizontal and vertical equal signs may be found on the same page (e.g., Add. MS 6783, f. 116, a page giving the solution of a pair of conjugate biquadratic equations by reduction). Occasionally (as on Add. MS 6782, f. 262) there is a long equal sign diagonally across the page, clearly for convenience.

10. The term is derived directly from Viète's *Isagoge*, Chapter V, "Concerning the Laws of Zetetics," Proposition 1. Viète describes the process as "transposition under opposite signs" and justifies it largely verbally.

11. Derived directly from Viète's *Isagoge*, Chapter V, Propositions I and II, "Hypobibasm is like the lowering of the power..." writes Viète and the derivation is from the verb ὑποβιβάζω, to lower. It means dividing throughout the equation by the unknown. Liddell and Scott give "carry over, transport". Again, a verbal explication is given by Viète in contrast to Harriot's (to us) crystal-clear symbolic demonstration.

12. Parabolismus is derived from Propositions II and III of Chapter V of Viète's *Isagoge* and means dividing throughout the equation by the same known quantity. The word comes from παραβάλλω (to apply) as above (p. xx).

Notes on Section Two

1. The processes covered in this Section and the two following are treated differently in the surviving manuscript papers. Canonical equations appear sporadically throughout the papers but there is only one block in which there is a systematic treatment in which the working is given and that is in Add. MS 6783, ff. 163–183, taken in the reverse order (as they have been incorrectly bound). The heading at the top of the pages is *On The Generation of Canonical Equations* and here such equations are "generated" by the product of binomial factors, their roots verified, and one or two terms are removed as in Sections 2, 3, and 4 of the *Praxis*. These pages are recommended by Torporley in the *Corrector Analiticus* for inclusion in a work on Harriot's algebra. F. 183 is typical and on this page it can be seen that what is separated into three separate Sections in the present work [i.e., the generation of Canonical equations, the removal of one or two terms and the designation of the (positive) root], is treated as a unity by Harriot himself in his own papers. (Cf. Add. MS 6784, f. 55 for a list of multiplications of binomial factors leading straight to Canonical equations without working.) Note the incompleteness of the treatment of Canonical Equations in the papers, e.g., ff. 176, 174 in which no working is given in removing a^3, a^2 and a respectively.

Sections 2, 4, and 5 refer to "Propositions" whereas Sections 3, 6 and the Numerical Exegesis refer to "Problems". The distinction between these, as seen by the seventeenth century lexicographer Macraelius is given as:

Propositio mathematicis est, quando proponitur in genere id, quod demonstrari debet. Et quidem si de subjecto aliquid demonstrandum, illa dicitur *theorema*: si autem aliquid est construendem in dato subjecto, dicitur *problema*, e.g., Theorema est, si oportet demonstrari, in omni triangulo tres angulos esse aequales duobos rectis. Sed problema est, quando quis conatur demonstrare, supra rectam lineam finitam posse strui triangulum aequilaterum. (p. 1150).

That is, in translation:

A proposition (in mathematicians' usage) is when something is proposed in general terms, which needs to be demonstrated. Now, if something needs to be *demonstrated* about some given, then that is called a theorem; but if something needs to be *constructed* upon some given, then that is a problem. For example, it is a theorem if one has to demonstrate that in every triangle the three angles are equal to two right angles. But it is a problem when someone tries to demonstrate that upon a finite straight line an equilateral triangle can be constructed.

2. Canonical equations are explained in Definitions 15, 16, and 17. Their form is not one in which an expression is equated to zero. The powers in equations are in descending order, the highest positive, so the "homogeneous term" may be negative. In the Harriot papers, however, the highest power may be a negative term.

3. "Originals" are conceived of as equations, not as identities, as they would be today. (See Definition 14 above.)

4. The convenience of Harriot's sign for multiplication is apparent here. The "original" corresponds to Proposition 2 of the *Praxis* (p. 16). The sign is used throughout the MS.

5. This is the first of eight cases in which the "original" would lead to an equation only by assuming negative or imaginary roots. None is included in what follows. Certain other products are also omitted, such as $(a + b)(a - c)$, implying one positive root, presumably because nothing intrinsically different is achieved by their inclusion.

6. The following is a table showing the correspondence between the "originals" and the canonical cubics derived from them in the present Section.

Original no.	Proposition in which canonical is derived
1	5
2	4
3	3
4	No corresponding canonical
5	8
6	6
7	7
8	No corresponding canonical
9, 10, 11	No corresponding canonical Completed after Note, (p. 27) in somewhat different form.

Note that there is an interchange of letters in some cases, but the signs correspond (Proposition 3 and original 3, Proposition 8 and original 5, Proposition 6 and original 6, and Proposition 7 and original 7).

7. Although it is not necessary to give cases which will not be used, all four cases are given here and below. Might this possibly reflect Warner's familiarity with the papers themselves or a possible draft by Harriot?

8. Cubics 9 and 10 have powers in ascending, not descending, order. Perhaps so that the right-hand side shall be positive? All three cubics 9, 10, and 11 are significant in connection with the solution of cubics in Section 6, Problems 12 and 13 (cf. Add. MS 6783, ff. 101–103). (These are some of the pages in the manuscripts in which cubic equations are solved.

9. This is the first occurrence of "equating to zero," but it appears only as a step on the way to canonical form, not as a final expression. More impressive than this is Add. MS 6783, f. 187 in which zero is treated as a calculative entity. On the

bottom right-hand corner, we read:

"$eee - 0e = +0 = +0$, $e - eee$

$$e = +0 \qquad\qquad\qquad e = +0$$
$$e = -0 \qquad\qquad\qquad e = +0$$
$$e = -0 \qquad\qquad\qquad e = -e$$
$$\overline{} \qquad\qquad\qquad \overline{}$$
$$a = e + r \qquad\qquad\qquad a = r - e$$

$$a = 0 + 2 = +2 \qquad\qquad a = +2 - 0 = +2$$
$$a = -0 + 2 = +2 \qquad\qquad a = +2 - 0 = +2$$
$$a = -0 + 2 = +2 \qquad\qquad a = +2 + 0 = +2$$

Hic est: $a = r$"

Note that on this manuscript sheet we not only have addition with zero, but also multiplication.

10. The method of "generating" equations is an excellent device for exhibiting the structure of the equation with respect to its roots. However, omission of the negative roots (as in the Proposition) would prevent Harriot from seeing the relations between roots and coefficients (what we call Symmetric Functions of the Roots). The display of the equations is such that these functions are only implicit. Nowhere in either *Praxis* or MS are the relationships stated, so we may only surmise that he was aware of such a connection. Add. MS 6783, f. 157 (in the British Library) has a statement about the sum of the roots of a biquadratic equation and all four roots are given, of which two are imaginary. Harriot's pages are inconsistent on such presentations, as the results given below clearly show. (See also J. Stedall, 2003, p. 268.) The best we can justifiably say is that with hindsight the way in which Harriot generated equations from multiplication of binomial factors displayed the connection. No evidence exists of Harriot's awareness (or otherwise). The same holds for the connection between the dimensions of the highest power and the number of roots of an equation.

The *Praxis* presents a system recognising only positive roots (except the intrusions on pp. 95 and 97); and throughout Section 4, in which roots of both Primary and Secondary Canonicals are given, care is taken to consider *positive roots only*. It is significant that the pages in the MS showing the generation of Canonical Equations (ff. 163–183) have algebra purporting to demonstrate that the Canonical Equations have only the positive roots and no others (by showing that the assumption of a further root leads to a contradiction). So, although there are many, many cases in the papers of recognition of negative roots, and some complex roots, it is impossible to assert that Harriot saw the connections that now seem appropriate.

Propositions 6 and 13 generate equations involving imaginary roots but in no case are factors considered which relate to negative or imaginary roots. The following table shows the actual roots of the canonical equations derived in this Section and the roots which are given in the text, in each Proposition.

Proposition no.	Actual roots	Roots given
1 (p. 16)	$b, -c$	b
2 (pp. 16–17)	b, c	b, c
3 (p. 17)	$b, -c, -d$	b
4 (pp. 17–18)	$b, c, -d$	b, c
5 (pp. 18–19)	b, c, d	b, c, d
6 (p. 19)	$b, \pm icd$	b
7 (pp. 19–20)	$-b, (a^2 = cd)$	$aa = cd$
8 (p. 20)	$b, (a^2 = cd)$	$aa = cd$
9 (pp. 20–21)	$b, -c, -d, -f$	b
10 (pp. 22–23)	$b, c, -d, -f$	b, c
11 (pp. 23–24)	$b, c, d, -f$	b, c, d
12 (pp. 24–25)	b, c, d, f	b, c, d, f
13 (p. 25)	$b, \pm 3\sqrt{} - cdf$	b
14 (p. 26)	$-b, (cdf = aaa)$	$cdf = aaa$
15 (p. 26)	$b, (cdf = aaa)$	$b, (cdf = aaa)$

11. Note that a is equated to both b and c.

12. Another example of the many in which $+$ and $-$ are treated as operators and directional signs (as is the case today).

13. This is the first case in which an imaginary root is ignored. Pages in the manuscripts exist in which imaginary or even complex roots are stated (Add. MS 6783, ff. 49, 156, 301). f. 49 and following pages lists a number of biquadratic equations together with solutions and, although negative roots appear, together with some imaginary ones, complex roots are not given. f. 156 shows the generation of seven canonical biquadratic equations. The first five give three roots, ignoring one negative root although one imaginary root is there. The last two derivations are not completed, simply putting $aa = bc$ and $aa = df$, in one case and $aa = bc$, $aa = -df$ in the other. f. 301 lists quadratic, cubic, and biquadratic equations with some roots and one biquadratic has four roots, two real and two complex, completely and correctly written out. This is the very equation from which the terms in x^2 and x^3 have been removed and which Warner refers to in Section 3 (see *Praxis*, p. 46). Also see S. P. Rigaud, *Miscellaneous works and correspondence of the Rev. James Bradley*, D. D., F.R.S., 2 vols (Oxford, 1832–1833). *Supplement....*) Harriot's manuscripts are inconsistent in including complex roots. Where he writes, for example, on the Rigaud facsimile

$$a = +1 + \sqrt{-32}$$
$$a = +1 - \sqrt{-32}$$

all that may be inferred is a sort of instrumental acceptance that the working and notation lead to this. Harriot's use of the word "noetic" in reference to imaginary roots (Add. MS 6783, f. 157) indicates a degree of recognition (at least at the time of writing this manuscript).

That Harriot was, at the very least, extremely close to "recognition" of complex roots is shown in his MS (6783, f. 58v) where he is "playing" with equations $a^4 = 16$ and $a^4 = -16$.

The situation with respect to complex and imaginary roots is mixed. In the hundreds of cubics and biquadratics listed with solutions, it is almost always the case (in the papers) that only real roots are given. Thus, for cubics, either one or three roots are normally stated and in the case of biquadratics, two or four. The MSS very often give negative roots when lists are presented (see, especially, Add. MS 6783, f. 49–57, 59–60). Such roots are included for polynomial equations with numerical coefficients. They are not given for the equations considered in Add. MS 6783, ff. 163–183 (see above).

Summary of results on 400 solved equations in manuscripts

1) 5 wrong.
2) 144 fully and correctly solved.
3) Special cases on Add. MS 6783, f. 49. All biquadratic:

  a) 2 give 4 real roots.
  b) 1 gives 4 imaginary roots.
  c) 1 gives 2 imaginaries, omits negative imaginaries.
  d) 4 gives 1 real, 1 imaginary, negatives omitted.
  e) 1 gives 2 real, 2 imaginary. (Negatives given).
  f) 1 ends "$aa = \pm \sqrt{-6}$" Before this "$aa = 4 - 4$".

4) 199 leave out complex roots.
5) 6 end with e.g., "$aa = -4$".
5a) 12 omit imaginaries.
6) 1 omits negative imaginary but gives $\sqrt{-4}$.
7) 7 cubics, coincident roots ignored.
8) 1 biquadratic ignores 1 real root.
9) 8 cubics omit 1 real root, not coincident with others.
10) On f. 53^v, coefficient of highest power $= 1$.
11) On f. 53^v, coefficient of highest power $= 1$. Root $= \dfrac{-3}{2}$
12) 1 biquadratic omits 2 real roots.
13) 1 cubic omits 2 real roots.
14) 2 biquadratics omit 2 real roots (irrational).
15) 1 quadratic omits 1 real root.

14. The following table shows in which Proposition canonical biquadratics are derived from their "originals".

Original number	Proposition in which canonical is derived
1 (p. 13)	12 (pp. 24–25)
2 (")	11 (pp. 23–24)
3 (p. 14)	9 (pp. 20–21)
4 (")	10 (pp. 22–23)
5 (")	No corresponding canonical
6 (")	15 (p. 26)
7 (")	14 (")
8 (")	13 (p. 25)
9 (")	No corresponding canonical

There is one interchange of letters in the original equation, in Proposition 9 and original 3. Biquadratics 10 to 18 on pp. 14–15 are not treated.

15. Paragraph 1 is inaccurate; the list of biquadratic originals on p. 15 of the *Praxis* goes from 10 to 18, the last one being:

$$\left. \begin{array}{l} bc + aa \\ df + aa \end{array} \right| \begin{array}{l} = bcdf + dfaa \\ + bcaa + aaaa \end{array}$$

There are, therefore, nine types and not eight.

Paragraph 2 in the original text of the *Praxis* mistakenly selected cubic no. 10 instead of no. 18.

The first and only reference to negative (i.e., "privative" roots) is contained in this note in Paragraph 2. The editor explains the omission of the "derivation of canonical equations from certain original equations" on the grounds that they can only be carried out by "the substitution of negative (privative) roots" and, therefore, "they are omitted insofar as they are useless". The interesting point emerging here is that it is not *existence* which is the issue but *usefulness*. Such a criterion is understandable in the light of the practical interests of both Harriot and Warner. What is not explained by this is the enormous number of negative roots which are found in the Harriot manuscript papers. Many of these occur in lists of equations and solutions given without showing the working which might seem to be exercises for students. However, in the block of pages referred to above, that is Add. MS 6783, ff. 183–163, there are cases in which he is generating equations which we would see as having negative roots (e.g., f. 183), in which he shows their positive roots by substitution and goes on to "prove" that there can be no others.

16. These "three special equations" are equivalent to original equations 9, 10, and 11 on p. 13.

17. The list of cubic "originals" given at the beginning of the Section, all were generated except those having only negative or imaginary roots. In a few cases (reciprocal cubics), the lettering was changed, e.g., original No. 5 is written:

$$\left. \begin{array}{l} aa - bc \\ a - d \end{array} \right| \qquad \text{but Proposition 8 has} \qquad \left. \begin{array}{l} a - b \\ aa - cd \end{array} \right|$$

although these clearly correspond to each other.

Of the biquadratic "originals", 1, 2, 3, and 4 correspond to Propositions 12, 11, 9, and 10, and 6, 7, and 8 correspond to Propositions 15, 14, and 13.

Notes on Section Three

1. In this Section Primary Canonical equations generated in Section 2 are reduced to Secondary Canonicals, that is, Primary Canonicals in which the next to highest degree term is missing (or two terms are missing), by equating the appropriate coefficient(s) to zero. The posited root is unchanged but what is not stated is that equating the appropriate coefficient(s) to zero inevitably implies a special relationship between the roots. At first glance, it might seem curious to devote a Section to this, in the light of the undoubtedly more sophisticated method of removing terms laid out in Section 6 (see below, pp. 87 *Praxis*). However, the equations to be reduced in Section 6 are not in the same Canonical form as in this Section but come under the heading of what the editor(s) will call Common Equations.

The relationship between the roots for a zero coefficient used in Section 3 will not be used in Section 4, which gives roots of primary and secondary Canonicals, in which only positive roots are demonstrated. (See table of correspondences between Sections 3 and 4 pp. 229–31)

2. Problem 1 demonstrates the procedure very simply. The roots of the Primary Canonical are b and $-c$, (the latter implicit but not recognised). When $b = c$, the equation becomes $aa = bb$, whose roots are, in our terms, $\pm b$.

Problem 1 is the only quadratic equation treated in this Section, surely because the equation deduced in Section 2, Proposition 2 might raise the problem of negative roots ($b = -c$).

3. The division line is used for the first time in a way that would be ambiguous for us at this point and twice more in Problem 3. It refers to all terms written above it and is repeated below (e.g., Problems 5 and 10).

4. Problems 6, 7, and 8 differ somewhat in form from the preceding ones and correspond to Section 4, Propositions 12, 10, and 11 respectively. It is the special form of these equations that lends itself to the different method.

Note item 9, p. 13 of the *Praxis* in which Primary Canonical $rrr - 3.rra + 3.raa - aaa = +qqq$ is generated from $r - a = q$, and which corresponds to the present case.

5. Note item 10, p. 13 in which Primary Canonical $rrr + 3.rra + 3.raa + aaa = +qqq$ is generated from $r + a = q$, and which corresponds to this equation.

6. Note item 11, p. 13 in which Primary Canonical $aaa - 3.raa + 3.rra = rrr = +qqq$ and which corresponds to this equation.

7. Note that there has been a sign reversal in denominators. The signs in front of the coefficients in the trinomial in Problem 17 are opposite to those in Problem 16.

8. The Note below by Warner (p. 46) refers back to Problems 16, 17, and 18 in which only one term is removed, the terms in aaa, aa and a respectively. Inspection shows that the reduction in Problem 17 is only a variant of Problem 16.

Problems 19–21 are also referred forward by him to Section 4, Propositions 35–37 in which positive roots are stated. (See below n. p. 91–2)

Problems 19, 20, and 21 are of particular interest as it is obvious that it is not possible to remove two terms from an equation by equating one coefficient to zero. Two independent relations are needed in these cases.

The most important thing, however, is the final sentence: "Since, however their reductions have been handed down most obscurely in manuscript they must await a more thorough enquiry."

In fact, Add. MS 6783, ff. 174, 173, 172, and 204 have completely worked-out reductions of the given biquadratic, correctly removing two terms in each case. ff. 174–172 are marked d. 10, 11, and 12 at the top left and f. 204 is marked d. 7.2°, showing that the last has been wrongly bound and should have been next to f. 177 (d. 7). Surely f. 204 comes very neatly after f. 177 because, having performed the removal of one term in the biquadratic, it removes two terms. f. 176 returns to removing one term from a different biquadratic.

The algebra in the four manuscript pages is more "dense" than appears at first sight. The steps are, however, simple in themselves and may easily be elucidated, although our own elucidation is unlikely to correspond to Harriot's thought processes. Harriot had great facility in symbolic thinking and may well have needed no more than is shown on the pages.

Two conditions are used, f is eliminated and a quadratic in d is obtained and solved in terms of b and c. f is then obtained, also in terms of b and c by substitution. After obtaining $d + f$ and df, the coefficient and homogene of the reduced equation is obtained, giving two real roots only.

It is very significant that, although the unreduced biquadratic on f. 204 is credited with four roots, the reduced form has only two roots attributed to it. This point is obscured in f. 174 to 2 in which the unreduced biquadratic is credited with only two roots to start with.

Elucidation of Add. MS 6783, f. 174 (fig. 1)

The biquadratic from which terms in aaa and aa are to be removed is:

$$aaaa - baaa + bcaa$$
$$- caaa - bdaa$$
$$+ daaa - cdaa + bcda$$
$$+ faaa - bfaa + bcfa$$
$$- cfaa - bdfa$$
$$+ dfaa - cdfa = -bcdf$$

$$\text{Put } b + c = d + f$$
$$\therefore b + c - d = f - - - - - -(1)$$

(The superscripts here and in the following do not indicate powers but are used by Harriot to help in the correct collection of terms.)

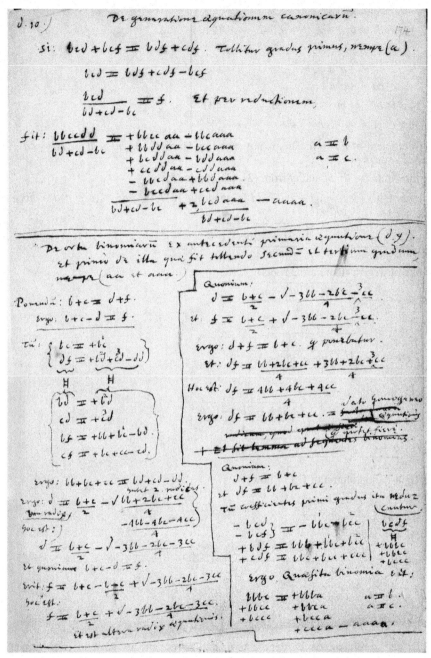

Figure 1. The British Library Board. All rights reserved. Add. Ms 6783, f. 174.

$$bc = +b^1c$$

(1) $x\ d$; $df = b^2d + c^3d - dd - - - -(2)$

$$\| \quad \|$$

$$bd = +b^2d$$
$$cd = +c^3d$$

(1) $x\ b$; $bf = +bb + b^1c - bd - - - - -(3)$

(1) $x\ c$; $cf = cb + cc - cd - - - - - - (4)$

{Coefficient of $aa = 0$. This is not stated by Harriot, i.e., $bc + df = bd + cd + bf + cf$ —(5).}

{Substitute for df, bf and cf from (2), (3), (4), in (5)}

$\therefore bb + bc + cc = bd + cd - dd$ [a quadratic in d, i.e., $d^2 - (b + c)d + bb + bc + cc = 0$

$$\therefore d = \frac{b+c}{2} \pm \sqrt{\frac{-3b^2 - 2bc - c^2}{4}} \text{ (for us)]}$$

Harriot writes: (there are 2 roots)

$$d = \frac{b+c}{2} - \sqrt{\frac{-bb + 2bc + 3cc}{4}}$$

$$- \frac{4bb - 4bc - 4cc}{4}$$

$$d = \frac{b+c}{2} - \sqrt{-\frac{3bb - 2bc - 3cc}{4}}$$

And since $b + c - d = f$

$$f = b + c - \frac{b-c}{2} + \sqrt{-\frac{3bb - 2bc - 3cc}{4}}$$

$$f = \frac{b+c}{2} + \sqrt{-\frac{3bb - 2bc - 3cc}{4}}$$

[Right-hand column of f. 174 now]

$$d = \text{as above}$$
and $f = \text{as above}$

$\therefore d + f = b + c$

And $df = \dfrac{bb + 2bc + cc}{4} + \dfrac{3bb + 2bc + 3cc}{4}$

$$df = \frac{4bb + 4bc + 4cc}{c}$$

$\therefore df = bb + bc + cc$

{To get coefficient of 1st degree term and homogeneous term in transformed equation: eliminate $d + f$ from original equation's coefficient of a and homogeneous term.}

$-bcd$

$$= -bc\,(d+f)$$

$-bcf$

$$\overset{1\ \ 2}{= -bc\,(b+c)} = -bbc - bcc$$

$$\overset{1\ \ 2}{+\,bdf = b\,(bb+bc+cc)} = bbb + bbc + bcc$$
$$+\,cdf = c\,(bb+bc+cc) = bbc + bcc + ccc$$

$$bcdf = bc\,(bb+bc+cc)$$
$$= bbbc + bbcc + bccc$$

So the sought binomial is:

$bbbc =$	$bbba$	$a = b$
$+\,bbcc$	$+bbca$	$a = c$
$+bccc$	$+bcca$	
	$+ccca\ -aaaa$	

{f. 204 is a little different but with very similar working.}

Two points of notation are to be noted. In the course of the working there are small numeral superscripts which, at first sight, look like indices. They are not, however, and turn out to correspond to similar terms (for example, bd has 2 above it) when there is "gathering up" to be done. The second point is that, whereas on f. 174 (d. 10) the expression under the square root sign in the solution to the quadratic equation is written in full, this is not pursued on d. 11 and d. 12 and Harriot writes exactly as we do today if thought outstrips writing-speed or we feel that the entire expression may be taken as understood.

It so happens that Wallis (*Algebra*, 1685 p. 151) states the results together with four roots for each reduced biquadratic and including the complex roots arising in Problem 19. These had been given without explanation in Add. MS 6783, f. 301. Of course, Wallis was perfectly well able to deduce all the roots for himself, and he must, at some time, have worked out the solutions or seen the papers perhaps in a copy owned by Pell. If it were the former, then this work would probably have been done in notes accompanying his study of the *Praxis*, read later when writing the *Algebra* when he might have assumed that the notes had been in the *Praxis* itself. It is very likely that he had seen the papers.

We cannot really speculate as to why Warner backed away from these solutions (perhaps fear of the square-root of a negative quantity?) as it would be historically improper to attempt to read his mind. It may very well have been genuine confusion since Harriot does not actually state the second condition in his working, he simply uses it.

9. The problem corresponding to this equation should be Number 15 but the signs of the coefficients are changed. All others in the list correspond exactly to what has gone before.

10. The Section ends with lists of canonical forms beautifully set out so that the pattern of their formation emerges clearly and so that the pattern may be extended indefinitely to those of a higher degree. For example, the first list (p. 25) might be extended to an equation of the sixth degree:

$$+ bbbbbc \qquad\qquad + bbbbba$$
$$+ bbbbcc \qquad\qquad + bbbbca$$
$$+ bbbccc \qquad\qquad + bbbcca$$
$$+ bbcccc \qquad\qquad + bbccca$$
$$+ bccccc \qquad\qquad + bcccca$$
$$\qquad\qquad\qquad\qquad + ccccca - aaaaaa$$

In three places we have "And so to infinity (in other cases) by the same method".

Notes on Section Four

1. See Note 1, Section 3 and adjoining table for some comparison of the present Section (and its relation to Sections 2 and 3).

2. As we would expect, only the positive roots are given and in some cases below a Lemma claims to prove that the roots obtained are the only possible ones satisfying the equation.

 The note on "privative" (i.e., negative) roots on p. 27 is validated. In fact, having declared that such roots are to be disregarded the text is not required to mention them again and this despite the fact that, below on pp. 95–97, two negative roots creep in (presumably by accident). Proposition 1 clearly does not consider the root arising from the factor $a + c$ in the equation which may be written (in modern notation)

$$(a - b)(a + b) = 0$$

3. The word "radix" is translated as "the root" sooner than "a root" in line with the work's assumptions about negative roots.

4. The Latin is "Ergo radici a posito b aequalis...". Therefore b is supposed equal to a. Thus b was "posited" as a root (or value of a) and is a "posited" root until it is proved to be an "explicated" root in the Proposition.

5. The equations whose roots are considered in this section correspond, on the whole, to the Canonical Equations of the previous Sections. (See fig. 229–31 for a table of correspondence). The method used speaks for itself and consists in substituting the positive root(s) for a, and demonstrating the equality. The present Section does satisfy the criteria given in Definition 8 for Poristic.

6. It is unfortunate that the alternative (positive) root is taken as c, when the negative root would be $-c$ arising from factor $a + c$. The same substitution is used by Harriot in the manuscript (see top right-hand corner of Add. MS 6783, f. 183). The word Lemma is not used by Harriot who treats the work on a particular equation as a unity.

7. a is put equal to d in the Lemma (cf. n5).

8. This is shown by inspection of Propositions 3 and 4 on pp. 53, 54 of the *Praxis*. The left-hand side should be:

$$+ aaa + baa + bca$$
$$+ caa - bda$$
$$- daa - cda$$

9. The notation has been modernised for ease of printing. It originally had connecting vertical equality signs.

The following table connects each Proposition in Section 4 with the corresponding Problem in Section 3, together with a comparison of the roots of equations recognised in Section 4 with the actual roots.

Number of Proposition in Fourth Section	Page	Number of Problem in Third Section	Page	Actual Roots or Equations giving these roots	Roots recognised in the Fourth Section
				Quadratics	
1	52	(Pr.) 1	29	$b, -c$	b
2	52	None	—	b, c	b, c
				Cubics	
3	53	(Pr.) 4	31	$-b, -c, d$	d
4	54	(Pr.) 3 (1)	30	$-b, c, d$	c, d
5	55	None	—	b, c, d	b, c, d
				Reduced Cubics	
6	57	(Sec.) 2	29	$b, c, -(b+c)$	b, c
7	57	(Sec.) 4	31	$-b, -c, b+c$	$b+c$
8	58	(Sec.) 3	30	$b, c, \dfrac{-bc}{b+c}$	b, c
9	59	(Sec.) 5	32	$-b, -c, \dfrac{bc}{b+c}$	$\dfrac{bc}{b+c}$
				Special Cubics	
10	59	(Pr.) 7	33	$c - b, \tfrac{1}{2}(-2b - c \pm i\sqrt{3}c)$	$c - b$
11	60	(Pr.) 8	34	$c + b, \tfrac{1}{2}(2b - c \pm i\sqrt{3}c)$	$c + b$
12	60	(Pr.) 6	33	$b - c, \tfrac{1}{2}(2b + c \pm i\sqrt{3}c)$	$b - c$

13	60	(Pr.) 8 (2)	34	$2b, \tfrac{1}{2}(b \pm i\sqrt{3b})$	$2b$
14	61	(Sec.) 7	33	$c - b, \tfrac{1}{2}\{(b - c) \pm \sqrt{3}i(b + c)\}$	$c - b$
15	61	(Sec.) 8	34	$c + b, \tfrac{1}{2}\{-(b + c) \pm i\sqrt{3}(b - c)\}$	$c + b$
16	61	(Sec.) 6	33	$b - c, \tfrac{1}{2}\{(c - b) \pm i\sqrt{3}(c + b)\}$	$b - c$
17	62	(Sec.) 8 (3)	34	$2b, -b, -b$	$2b$
				Reciprocal Cubics	
18	62	None	—	$b, a^2 + cd = 0$	b
19	62	None	—	$c, -c, -b$	c
20	62	None	—	$b, c, -c$	b, c
				Biquadratics	
21	63	(Pr.) 12	39	$-b, -c, -d, f$	f
22	64	(Pr.) 9	34	$b, c, d, -f$	b, c, d
23	65	(Pr.) 15	42	$b, c, -d, -f$	b, c
24	67	None	—	b, c, d, f	b, c, d, f
				Reduced Biquadratics	
25	68	(Sec.) 9	34	$b, c, d, -(b + c + d)$	b, c, d
26	69	(Sec.) 10	36	$b, c, d, -\dfrac{bc+bd+cd}{b+c+d}$	b, c, d
27	70	(Sec.) 11	37	$b, c, d, -\dfrac{bcd}{bc+bd+cd}$	b, c, d
28	71	(Sec.) 12	39	$-b, -c, -d, b + c + d$	$b + c + d$
29	71	(Sec.) 13	40	$-b, -c, -d, \dfrac{bc+bd+cd}{b+c+d}$	$\dfrac{bc+bd+cd}{b+c+d}$

(Continued)

Number of Proposition in Fourth Section	Page	Number of Problem in Third Section	Page	Actual Roots or Equations giving these roots	Roots recognised in the Fourth Section
30	72	(Sec.) 14	41	$-b, -c, -d, \dfrac{bcd}{bc+bd+cd}$	$\dfrac{bcd}{bc+bd+cd}$
31	72	(Sec.) 15	42	$b, c, -d, -(b+c-d)$	b, c
32	73	(Sec.) 16	43	$b, c, -d, -\dfrac{bc-bd-cd}{b+c-d}$	b, c
33	74	(Sec.) 17	44	$b, c, -d, -\dfrac{bd+cd-bc}{d-b-c}$	b, c
34	74	(Sec.) 18	44	$b, c, -d, \dfrac{bcd}{bd+cd-bc}$	b, c
35	75	(Sec.) 19	45	$b, c, \tfrac{1}{2}\left(-b, -c \pm i\sqrt{3b^2+3c^2-2bc}\right)$	b, c
36	76	(Sec.) 20	45	$b, c, \tfrac{1}{2}\{-bc\,(b+c) \pm ibc\sqrt{3b^2+3c^2+2bc}\}$	
37	76	(Sec.) 21	46	$b, c, -b, -c$	b, c
				Reciprocal Biquadratics	
38	76	None	—	$b, a^3 + cdf = 0$	b
39	77	None	—	$c, a^3 + (b+c)\,a^2 + c(b+c)a + bc^2 = 0$	c
40	77	None	—	$b, c, \tfrac{1}{2}(-c \pm i\sqrt{3c})$	b, c

Key

Pr.	=	Primary Canonical
Sec.	=	Secondary Canonical
(1)	=	Letters interchanged
(2)	=	Special case
(3)	=	Special case

Notes on Section Five

1. In this Section the numbers of roots of commonly occurring (ordinary) equations are to be determined by comparison with certain Canonical Equations whose (positive) roots have already been determined in Section 4. Many years later Wallis was to describe "ordinary equations" as "such as usually occur" (*Algebra*, 1685, p. 155). The work in Section 5 corresponds to certain of Harriot's MS pages labelled "*e*" in the top left-hand corner and headed "De resolutione aequationû per reductionem" (on the Solution of Equations by Reduction) linking them with the work in Section 6 in which cubic and biquadratic equations have certain terms removed. In Torporley's *Summary*, the passage corresponding to Section 5 is between 44ᵛ and 46ᵛ. What is clear is that Torporley's copy shows a draft in which all the work on cubics comes together (to be followed by biquadratics). There is no separate section corresponding to Section 5 of the *Praxis*, no mention of equipollence and no definition such as is given at the beginning of Section 5.

2. The word "equipollent" is defined below, unfortunately ambivalently. The first sentence gives the impression that equipollence means that the equations are of the same degree and are similarly affected and coefficients and homogeneous terms are in agreement with each other as far as "greater than" and "less than" are concerned. The equations then have the same number of roots. The second part of the paragraph, which is more verbose, implies more than this but with no specific reference to the actual relationships between the coefficients and the homogeneous terms of the Canonical and ordinary equations which must obtain in order that the comparison may yield results and which are considered below. Wallis reads more into the Definition than is, perhaps, justified (Ref. Wallis, *Algebra*, 1685). Wallis writes:

"And he shews how (by comparing such with his Canonicks duly chosen) to determine the Number of Roots in such Equations; How many are Real (and not only Imaginary), and how many of those be Affirmative.

For which he lays this general ground; That every Common Equation hath the same Number of Roots, (and so affected; with its respective Canonick, like Graduated, like Affected, and duly Qualified.

And those he calls *Duly Qualified*; when they are so qualified as that every of its known parts (that is, all the *Coefficients*, and the *Absolute Quantity*,) duly compared, (that is, each of them being Divided by such a Number as is the Number of Members in the respective part of the Canonick; and then advanced to such a Degree that they both attain the same number of Dimensions;) the parts (so advanced and mutually compared,) are respectively Equal, or Greater, or Less, in the one Equation, as they are in the other."

Wallis does not use the word "equipollent."

3. The six Propositions in the Section are interspersed with seven Lemmas, proving certain inequalities, and all the Lemmas but this first one are used below.

Lemma 1

The working is elementary and rests on the unstated assumption that $p > q$. In the first line we have changed fecetur to secetur.

The following table shows the correspondence between the Lemmas in the *Praxis* and the corresponding pages in the manuscripts in which they appear. Column three shows where each Lemma is used in the *Praxis*.

	MS	Use in Praxis
Lemma 1	None Corresponding	Not used
Lemma 2	Stated, 6783, f. 106r Proved, 6783, f. 106v	Lemmas 5, 6, 7
Lemma 3	Stated 6783, f. 106r Proved, 6783, f. 106v	Lemmas 5 and 7
Lemma 4	Proved, 6783, f. 106 (e. 8)	Proposition 1
Lemma 5	Proved, f. 107 (e. 9)	Propositions 2 and 4
Lemma 6	Proved, f. 185 (e. 28)	Proposition 5
Lemma 7	Proved, f. 184 (e. 29)	Proposition 5

All the manuscript pages referred to here relate to the inequalities seen in Section 5 of the *Praxis*. Particularly noteworthy are the inequality signs used from now on, which differ from those given earlier (Section 1, n. 9).

4. *Lemma 3*

The proof is similar to that of Lemma 2. The Lemma is stated on Add. MS 6783, f. 107r and it is proved f. 106v.

5. Propositions 1, 2, and 3 all refer to the ordinary equation: $aaa - 3.bba + 2.ccc = 0$ when $c \gtreqless b$ respectively. Proposition 1 compares the ordinary equation with the Canonical one:

$$aaa - 3.rqa = rrr + qqq$$

Lemma 4 shows that, for the Canonical Equation,

$$\frac{rrr + qqq \mid rq}{\frac{rrr + qqq \mid rq}{4} \mid rq} > \quad, \text{i.e.,} \quad \left(\frac{homogeneous\ term}{2}\right)^2 \quad > \quad \left(\frac{coefficient}{3}\right)^3$$

In the ordinary equation, $b < c$, so $c^6 > b^6$.

In this equation, too, therefore, $\left(\dfrac{homogeneous\ term}{2}\right)^2 > \left(\dfrac{coefficient}{3}\right)^3$
(ignoring signs).

So the two equations are equipollent and have an equal number of roots. Hence, by Section 4, Prop. 14 (it should be 15), the ordinary equation has one root. That is the argument in the *Praxis*.

Now, equation $aaa - 3.bba = +2.ccc$ is a special case of the general cubic: $x^3 + ax + b = 0$ (in modern terms).

Its solution, according to Cardano's method is:

$$x = \sqrt[3]{\left\{-\frac{1}{2}b + \sqrt{\frac{b^2}{4} + \frac{a^3}{27}}\right\}} + \sqrt[3]{\left\{-\frac{1}{2}b - \sqrt{\frac{b^2}{4} + \frac{a^3}{27}}\right\}}$$

This cubic has 3 roots, all real or 1 real and 2 complex (conjugate).

The expression $\frac{b^2}{4} + \frac{a^3}{27} = \Delta$ is called the discriminant of the cubic as it "discriminates" between the natures of the roots, according to its sign.

$\Delta > 0$ gives three real roots
$\Delta = 0$ gives two equal roots
$\Delta < 0$ gives one real and two complex roots.

The ordinary equation may be written:

$$a^3 + (-3b^2a) + (-2c^3) = 0$$

Hence, $\Delta_1 = \dfrac{4c}{4} + \dfrac{(-3b^2)^3}{27}$ (if its discriminant is Δ_1)

$$= c^6 - b^6$$

Similarly, for the Canonical equation,

$$\Delta_2 = \frac{(r^3 + q^3)^2}{4} - r^3q^3 \text{ whose discriminant is } \Delta_2).$$

What is being asserted in Section 5 about "equipollence" has a certain resonance with the discriminant (of the cubic, see above) because $c \gtrless b$ (that is, $c^6 \gtrless b^6$) for the ordinary equation is compared with the corresponding relationship for the Canonical.

Modern Theory of Equations in considering the roots of $x^3 + ax + b = 0$ considers whether $\frac{b^2}{4} + \frac{a^3}{27} \gtrless 0$ in order to determine the number of real roots. The question in the *Praxis* is whether

$$\left(\frac{\text{homogeneous}}{2}\right)^2 \gtrless \left(\frac{\text{coefficient}}{3}\right)^3$$

is true similarly for the ordinary and Canonical equations. In the case of the ordinary equation,

$$\left(\frac{\text{homogeneous}}{2}\right)^2 = c^6 \text{ and } \left(\frac{\text{coefficient}}{3}\right)^3 = b^6$$

So, in effect, Warner is comparing $c^6 - b^6$ with the corresponding expression for the Canonical, the number of whose (positive) roots had been determined in Section 4. So $c^6 - b^6$ is a kind of "discriminant" as its sign compared with that of the corresponding expression for the Canonical, determines the number of positive roots.

The idea is a clever one, certainly the use of the inequalities and it is not wrong within the given parameters of an algebra with only positive roots. Remember Lagrange's excitement at reading this Section, especially in relation to the passage giving the condition for these roots. (See p. 14)

There are two sheets of paper among the Tanner Papers in the University of Liverpool containing a proof that the conditions set out in Section 5 are necessary but not sufficient (Tanner Papers, BOX 50).

We recall that Descartes' Rule of Signs uses the words "can have," in recognition of the possibility of complex roots (*The Geometry of René Descartes*, Dover Publications, New York, 1954, p. 160).

Unfortunately, little resembling the Propositions of this Section has yet to come to light in the papers, and what little there is does not contain the word "equipollent". Particularly significant is the fact that the Torporley copy does not refer to equipollence.

6. *Lemma 4*

This Lemma, unlike what we find in f. 106 of the MS, is established synthetically. f. 106r proves the same result by analysis i.e., it starts by assuming what is to be proved as if it were the case and ends with what is known to be true (Add. MS 6783, f. 106).

7. The same reasoning applies as in n. 6 above.

Proposition 2

The ordinary equation is compared with the Canonical one labelled in the papers as Elipticus.

$cccccc < bbbbbb$ means that, for the common equation

$$\frac{(homogeneous\ term)^2}{2} < \frac{(coefficient)^3}{3}$$

The same inequality holds for the Canonical Equation following Lemma 5, to be proved below (ignoring signs). The Proposition follows.

8. *Lemma 5*

Nothing could be more lucid that the demonstration presented, using the same "algebraic synthetic" process as before. f. 107 performs the demonstration analytically, and ends in a simple identity. Thus Warner is better than Harriot here!

9. *Proposition 5*

Wallis describes Section 5 of the *Praxis* in his *Algebra* 1685 pp. 155–159 without criticism but with a caution for his readers after giving Descartes' Rule of Signs.

"But this rule must at least be taken with this *Caution*; That the Roots be Real, not only Imaginary. For as to Imaginary Roots, there may be yet an uncertainty.

But how many of these be Real, and how many but Imaginary; will depend upon that other condition of *Harriot's* Rule; *viz.* That the compared Equations be *duly qualified* as to the Equality, Majority, or Minority of their respective parts.

As to the former of these, we have *Descartes'* concurrence, (but without the caution interposed, which is a defect: Of the latter, (if I do not mis-remember) he is wholly silent" (*Algebra*, 1685, p. 158).

In Proposition 5, the mathematics given is not valid unless the equation is:

$$aaa - 3.baa + 3.cca == +ddd$$

as Wallis had carefully changed it to on p. 157!

For, $bb > cc$ in the ordinary equation as printed means that:

$$\left(\frac{\text{coefficient of } a^2}{3} \right)^2 > \text{coefficient of } a.$$

However, Lemma 6 proves that:

$$\left(\frac{p + q + r}{3} \right)^2 > \frac{pq + pr + qr}{3}$$

which would mean that, in the ordinary equation:

$$\left(\frac{\text{coefficient of } a^2}{3} \right)^2 > \frac{\text{coefficient of } a.}{3}$$

Thus, the coefficient of a in the ordinary equation needs to be $+ 3.cc$ for the correspondence to give the desired result. This typographical error in Proposition 5 is not noted in the original list of Errata. We may note also that Edmund Halley (1646–1742) investigated this equation geometrically and, by consideration of the equation

$$z^3 - bz^2 + pz - q = 0$$

he concluded that "Also Prop. 5, Sect. 5, of our countryman Harriot's Ars Analytica, and Prob. 18 of Vieta's Numer. Potest. Resol. is hardly founded" [E. Halley, "On the Numbers and Limits of the Roots of Cubic and Biquadratic Equations," *Phil. Trans. Roy. Soc.* (translated), 1686–1687, 16, 395–407, esp. 398].

Buried among other pages, a page of the MS is to be found (Add. MS 4394, f. 392r) in the Warner manuscripts which has a possible connection with Section 5.

d. 4 is written at the top and the heading:

De generatione aequationum canonicorum, Tho. Harioti (On the Generation of Canonical Equations by Thomas Harriot)

Lower down, is written:

e. 10) *De resolutione aequationum per reductionem* (On the Solution of Equations by Reduction)

d. 4 and e. 10 are the signifiers which Harriot wrote at the top left-hand side of his manuscript pages to indicate to himself which topic he was concerned with. Certainly, the page connects with Harriot's d. 4 and e. 10 and attempts to make connections between the ordinary equation reduced in Section 6, Problem 13 and its corresponding Canonical Equation. No inequalities are to be found and no clear conclusion is arrived at.

10. *Lemma 6*

The page of the manuscripts corresponding to this clearly demonstrated Lemma (proved by "algebraic synthesis") is Add. MS 6783, f. 185 (e. 28) and the working there is equally clear although the proof given is analytical. Harriot begins (as before) by assuming what is wanted as given and ends with an identity.

11. *Lemma 7*

The situation is similar to that in Lemma 6 above. The corresponding manuscript page is Add. MS 6783, f. 184 (e. 29). Pages e. 8, 9, 28, 29, and e. 10 clearly go together. Lemmata 1 and 2 of e. 8 are referred to on e. 28 and e. 29 and the page (Add. MS 4394, f. 392), found among the Warner papers which resonates with this Section, refers to e. 10. Taken together, these pages make it more than probable that Warner had seen either these particular pages or developments upon them written by Harriot himself.

12. Wallis (p. 158) states that the inequality used in Proposition 6 (like the others) was proved by Harriot, but such a proof is not in the *Praxis* and has, so far, not come to light in the papers.

Notes on Section Six

1. The bulk of the Section (Problems 1–11 and 14–34) is concerned with removing the term of second highest degree from cubic and biquadratic equations. Solutions are not normally given, although the "reduction" has prepared them for solution. Only Problems 12 and 13 go further than removal of a term (with a corresponding change of root) and provide solutions. Yet the Harriot manuscripts provide many examples of cubics and biquadratics fully solved. As we have already seen, it seems possible that Harriot intended to have two separate sections, one on cubics and another on biquadratics.

2. Pp. 87 and 88 give rather simplified versions of the work carried out by Harriot in Add. MS 6783, ff. 163 and 164.

3. Compare Definition of equipollent above, p. 78 numbered p. 72.

4. *Problem 1*

Add. MS 6783, f. 91 has an attempt at solving an equation of this type and one solution is given. The page may have been intended as "waste" however, as it seems that Harriot is "playing" with $a - r = e$.

f. 198 does solve:
$$ccd = -baa + aaa$$
$$2xxx = -3.raa + aaa$$

The right hand column gives the equivalent of the second substitution in the *Praxis* and the three possibilities are worked through. The reduced equation is described as absurd ("absurdum") and the working, despite different letters being used, is remarkably similar.

Viète's *De aequationum recognitione et emendatione* (in *Opera Mathematica*, ed. Schooten, 1646, p. 130, II) uses a substitution like that given in the *Praxis*.

From the point of view of generality we may note that Problems 1, 2 and 3 deal with three out of four possible cases

$$aaa + 3.baa == -ccc \text{ is missing.}$$

5. *Lemma*

The reasoning rests on b, c, d being positive.

6. In Harriot's papers Add. MS 6783, f. 67r(e) has a solution to the equation
$$2xxx = +3raa + aaa \text{ by reduction and f. 197 (e. 16) solves}$$
$$ccd = +baa + aaa$$
$$2xxx = +3raa + aaa$$
also giving numerical examples.

Viète, *op. cit.*, p. 130, I corresponds to Problem 2.

7. *Problem 3*

In Harriot's papers, Add. MS 6783, f. 65r(e) has a solution of an equation of the same type: $2xxx = 3.raa - aaa$
f. 196(e17) has

$$ccd = +baa - aaa$$

$2xxx = +3.raa - aaa$, together with numerical examples. Viète, *op. cit.*, p. 130 presents the transformation but, like the Harriot papers, gives the equation in a form in which the coefficient of a^3 is negative. Presumably the editor(s) of the *Praxis* wished to simplify the situation for their readers. Alternatively, the editor(s) were not afraid of negative terms occupying the entire right hand side of the equation. Viète's work includes both substitutions.

8. *Problem 4*

Add. MS 6783, f. 195r(e. 18) deals with $ffg = + cda + b.aa + aaa$
$2xxx = + ppa + 3.raa + aaa$

There are numerical examples, some with three roots, others with only one.

The corresponding statement in Viète, also with a single substitution, is given, *op. cit.*, p. 130 (under second heading).

9. *Problem 5*

Add. MS 6783, f. 194(e. 19) solves the type: $ffg = -cda + baa - aaa$
$2xxx = -ppa + 3.raa - aaa$

The numerical example gives three roots.

The corresponding statement in Viète is given, *op. cit.*, p. 131, VII. Again, the editor(s) have arranged the equation so that the coefficient of a^3 is positive. Viète gives only one substitution, the one corresponding to the second one in the *Praxis*.

10. *Problem 6*

Add. MS 6783, f. 193(e. 20) deals with $ffg = -cda + baa + aaa$
$2xxx = -ppa + 3.raa + aaa$

General considerations are followed by numerical examples in which four equations each have 3 roots and one has one root only.

The Viète equivalent in *op. cit.*, p. 131 uses the same (single) substitution.

11. *Problem 7*

Add. MS 6783, f. 192(e. 21) deals with $ffg = +cda + baa - aaa$
$2xxx = +ppa + 3.raa - aaa$

The four equations that are solved are each given three roots.

The Viète equivalent is to be found in *op. cit.*, p. 131. The form of the equation in the *Praxis* is different (as above) from the MS and from Viète, who actually performs only the substitution corresponding to the second given here.

12. *Problem 8*

Add. MS 6783, f. 191(e. 22) solves the type: $ffg = -cda - baa + aaa$
$$2xxx = -ppa - 3.raa + aaa$$

and two solved numerical equations are each given 3 roots.

A comment at the bottom of the page relating to another equation calls it: "Hyperbolica aequatio".

The Viète equivalent is *op. cit.*, p. 131, IV. Only the first substitution in the *Praxis* is given by Viète.

13. *Problem 9*

Add. MS 6783, f. 190(e. 23) solves the type: $ffg = +cda - baa - aaa$
$$2xxx = +ppa - 3.raa - aaa$$

giving three roots to the cubic.

The Viète equivalent is *op. cit.*, p. 131, and only one substitution is given. What is particularly noteworthy in the *Praxis* is the negative root implied by the first substitution,

$$a = -e - b.$$

14. *Problem 10*

Add. MS 6783, f. 189(e. 24) solves the equation of type: $ffg = + cda - baa + aaa$
$$2xxx = + ppa - 3.raa + aaa$$

and two equations that are solved are given only one root each.

The exact equivalent in Viète is *op. cit.*, p. 131.

f. 187 is of considerable interest, as it solves five cubics of the present pattern. Three roots are given in each case (all are positive).

However, the main interest lies in the use of zero as a calculable quantity.

He writes, for example, $a = 0 + 2 = +2$
$$e = -0, e = +0$$

Significantly, he writes near the bottom of the right-hand column,

$$eee - 0e = +0 = +0, e - eee$$

Thus, there is multiplication as well as addition and subtraction with zero.

15. *Problem 11*

Add. MS 6783, f. 188(e. 25), which has no numerical cases, Harriot deals with:

$ffg = -cda - baa - aaa$
$2xxx = -ppa - 3.raa - aaa$

There is no equivalent in Viète. As on p. 95 of the *Praxis*, a negative root is implied by the substitution $a = -e - b$.

Problem 11 is the last of the cubics which are reduced without going forward to a solution.

16. Problem 12 is a departure from the previous Problems not only methodologically but also because it goes on to a solution, albeit giving one root only.

The substitution all but replicates that of Viète by putting

$$a === \frac{ee - bb}{e}, \text{ where Viète had:}$$

$$\frac{\text{"B planum} - \text{Equad}}{\text{E}} \text{ erit A", that is, (in modern notation)}$$

$$A = \frac{B^2 - E^2}{E}$$

The substitution is equivalent to putting a equal to the difference between the extremes of three terms in continued proportion.

All is clear down to:

$$+eeeeee - 2.ccceee + cccccc$$

$$= +cccccc + bbbbbb$$

Following this, $eee = ccc + \sqrt{ccccc + bbbbbb}$ but, unlike f. 102(e. 5) (the corresponding manuscript page) the other possibility, that the left-hand side may be the square of $ccc - eee$, is not considered. It is assumed that there will only be one root.

The Corollary uses three quantities in continued proportion and, in modern notation and printed without error (see Errata for original errors), would read:

$c^3 + \sqrt{(c^6 + b^6)}, b^3, -c^3 + \sqrt{(c^6 + b^6)}$, are in continued proportion,

$\sqrt[3]{c^3 + \sqrt{c^6 + b^6}}, b, \sqrt[3]{-c^3 + \sqrt{c^6 + b^6}}$ are also in continued proportion.

But $e = \sqrt[3]{c^3 + \sqrt{c^6 + c^6}}$

Therefore, $\dfrac{b^2}{e} = \sqrt[3]{-c^3 + \sqrt{c^6 + b^6}}$

Therefore, $a = \dfrac{e^2 - b^2}{e}$

$$= e - \frac{b^2}{e}$$

$$= \sqrt[3]{c^3 + \sqrt{c^6 + b^6}} - \sqrt[3]{-c^3 + \sqrt{c^6 + b^6}}$$

The solution corresponds to that of Cardano, which may easily be checked

If $x^3 + ax = N$, $x = \sqrt[3]{\left(\dfrac{a}{3}\right)^3 + \left(\dfrac{N}{2}\right)^2 + N} - \sqrt[3]{\left(\dfrac{a}{3}\right)^3 + \left(\dfrac{N}{2}\right)^2 - N}$

The three numerical equations are given 1 root each, the other pairs (conjugate complex numbers) are:

$$-1 \pm 3i, \; -1 \pm 2i \quad \sqrt{3}, \; \frac{-1 \pm 3i\sqrt{3}}{2}$$

Add. MS 6783, f. 102 corresponds to Problem 12. (See fig. 2)

Although the substitution is the same as that of Viète, the work is carried out in Harriot's unique methodology, using Canonical equations generated by binomial factors. The Canonical equation on f. 102(e. 4) is:

$qqq - rrr = +3.qra + aaa$. He writes $a = q - r$ and, on f. 101 he had generated the Canonical from this equation. He had generated $rrr - qqq = +3qra + aaa$ using $a = r - q$.

He writes: $a = q - r$

Let: $r + a = q$

Then: $a = q - r$

Therefore:

$$\left.\begin{array}{c} r+a \\ r+a \\ r+a \end{array}\right| = rrr + \underbrace{3rra + 3raa}_{\parallel} + aaa = qqq$$

$$\left.\begin{array}{c} +r \\ 3ra \end{array}\right| + \left.\begin{array}{c} a \\ 3ra \end{array}\right| = \left.\begin{array}{c} +q \\ 3ra \end{array}\right| = 3qra$$

Therefore: $3qra + aaa = qqq - rrr$ or:

$qqq - rrr = +3qra + aaa$, as above $a = q - r$

(cf. *Praxis*, Section 4, Proposition 14 and 16, p. 61).

The most general form of the equation heads the page (f. 102)

$ggh = +dfa + aaa$

The next line gives the "ordinary form":

$2ccc = +3.bba + aaa$

Only then does Harriot write the corresponding Canonical equation:

$qqq - rrr = +3\,qra + aaa \qquad a = q - r$

The working resembles, but is not identical to that in Problem 12 of the *Praxis*:

Let $e = q$, $\dfrac{b^2}{e} = r$ (in modern notation)

Note that the implication is that the three quantities in continued proportion are $e, b, b^2/e$.

Both the *Praxis* and Viète had written $a = e - \dfrac{b^2}{e}$ or its equivalent. Comparison of the homogeneous terms of the ordinary and Canonical equations gives

$e^3 - b^6/e^3 = 2c^3$

$$\Rightarrow e^6 - b^6 = 2c^3 e^3$$

$$\Rightarrow e^6 - 2c^3 e^3 = b^6$$

$$\Rightarrow e^6 - 2c^3 e^3 + c^6 = b^6 + c^6$$

1)
$$e^3 - c^3 = \sqrt{(b^6 + c^6)}$$
$$e^3 = \sqrt{(b^6 + c^6)} + c^3 = q^3 \text{ (from } e = q)$$
$$e = \sqrt[3]{(\sqrt{(b^6 + c^6)} + c^3)} = q$$

2)
$$c^3 - e^3 = \sqrt{(b^6 + c^6)}$$
$$=> c^3 - \sqrt{(b^6 + c^6)} = e^3$$

In handwriting next to this, Harriot has written: "But $c^3 \not> \sqrt{b^6 + c^6}$ etc. There-fore there is no second root unless it is negative. In that case it will be $-r^3$. As is shown below." Next, the other possibility is explored, i.e., put $e = r$ and $\dfrac{b^2}{e} = q$.

$$\frac{b^6}{e^3} - e^3 = 2c^3 \text{ and } b^6 - e^6 = 2c^3 e^3$$

$$\Rightarrow b^6 = 2c^3 e^3 + e^6$$

$$\Rightarrow b^6 + c^6 = 2c^3 e^3 + e^6 + c^6$$

$$\Rightarrow \sqrt{(b^6 + c^6)} = c^3 + e^3$$

$$\Rightarrow \sqrt{(b^6 + c^6)} - c^3 = e^3 = r^3$$

$$\Rightarrow \sqrt[3]{(\sqrt{(b^6 + c^6)} - c^3)} = e = r$$

Following this, Harriot writes: "From this it appears that $-r^3$ is equal to the second root above and all the solutions may be formed from those two roots."

The lay-out of the page, especially the way in which the handwriting is carefully "lined-off" suggests that both pieces of explanation were inserted only after the algebra had been done. After all, when he wrote the first piece, he must have already known what came below. And Harriot's complete solution, including the adventitious (or ordinary) equation on the top line are written below.

He writes:

$$a = q - r$$

$$a = \sqrt[3]{(\sqrt{(b^6 + c^6)} + c^3)} - \sqrt[3]{(\sqrt{(b^6 + c^6)} - c^3)}$$

$$\text{or } a = \sqrt[3]{(\sqrt{(b^6 + c^6)} + c^3)} - \frac{b^2}{\sqrt[3]{(\sqrt{(b^6 + c^6)} + c^3)}} \quad \text{for } \frac{b^2}{q} = r$$

Right at the bottom of the page are the same three equations as are given on p. 99 of the *Praxis*. (There are other pages, ff. 402r and (e. 7)v and f. 411 with further attempts but these may well be "waste".)

The similarity of the substitution to that of Viète renders it likely that Viète's work at least sparked off Harriot's symbolic solution of the cubic (the first to be done in this manner). Viète writes:

17. The following three numerical equations use "$\sqrt{}$ 3" for "cube root of". Such notation occurs only in the *Praxis*, and the papers use a number of others for different roots, e.g.,

$\sqrt{}$ $(\sqrt{}$ 7 + $\sqrt{}$ 3) (modern) denoted by $\quad \sqrt{} , \sqrt{}$ 7 + $\sqrt{}$ 3 $\qquad$ (6783, f. 6r)

$\sqrt[3]{}$ 26 + $\sqrt{}$ 675 (modern) $\qquad\qquad \sqrt{}$r.26 + $\sqrt{}$ 675 $\qquad$ (6783, f. 72r)

$\sqrt[3]{}$ (modern) $\qquad\qquad\qquad\qquad \sqrt[3]{}$ $\qquad\qquad$ (6783, f. 1r)

$\sqrt[4]{36}$ $\qquad\qquad\qquad\qquad\qquad$ $\wedge\!\!\!\sqrt{36}$ $\qquad\qquad$ (6783, f. 4r)

18. Problem 13 is the second departure from the previous Problems and is very similar to Problem 12. The substitution, $a = \frac{ee+bb}{e}$ follows Viète exactly (*De Aequationum Recognitione, et Emendatione*, in *Opera Mathematica*, 1646, p. 150), and, again, the extremes of quantities in continuous proportion are involved. Three cases are taken $c \gtrless b$, necessitated by the fact that, in the solution, the quantity under the square-root sign is $cccccc - bbbbbb$.

In the third case, $cccccc < bbbbbb$, so that the reduced equation involves the square root of a negative quantity. He calls the equation "impossible" because $\sqrt{-ddddd}$ is "inexpressible", i.e., cannot be evaluated in a numerical example. The rest of the working is clear. Problem 13 does not use the signs for "greater than" and "less than" but words. It is also noteworthy that the possibility

$$ccc = eee + \sqrt{} (cccccc - bbbbbb)$$

is not considered.

Finally, note the negative term on its own in the line

$$eeeeee - 2.cccceee = -bbbbbb$$

Again, the manuscript equivalent, f. 103, is more complicated than the *Praxis*. (See fig. 3)

The Canonical equation $qqq + rrr = -3qra + aaa, a = q + r$ had been generated on f. 101 (cf. Section 4, Proposition 15, p. 61 of the *Praxis*).
Harriot writes on f. 101: $\quad$ Let $a - r = q$
$$\qquad\qquad\qquad\qquad \text{Then } a = q + r$$
$$\qquad\qquad\qquad\qquad \text{And} - a + r = -q$$

Therefore: $\qquad$ $\begin{vmatrix} a - r \\ a - r \\ a - r \end{vmatrix} = aaa - \underbrace{3.raa + 3.rra}_{//} - rrr = qqq$

$\qquad\qquad\qquad$ $\begin{vmatrix} -a \\ 3ra \end{vmatrix} + \begin{vmatrix} r \\ 3ra \end{vmatrix} = \begin{vmatrix} -q \\ 3ra \end{vmatrix} = -3qra$

Therefore: $\qquad qqq + rrr = -3qra + aaa \qquad a = r + q$
Let $\qquad\qquad a = r + q$

Therefore: $qqq + rrr = \dfrac{-3qr}{r+q} + \dfrac{\begin{array}{c} r+q \\ r+q \\ r+q \end{array}}{} = \begin{array}{l} rrr + 3rrq + 3rqq \\ \quad - 3rrq - 3rqq \\ \qquad + qqq \end{array}$ And so it is

f. 103 is headed: $ggh = -dfa + aaa$

or $2ccc = -3bba + aaa$

$\quad\quad\quad\quad\quad\quad\quad\quad \| \quad\quad\quad\quad\quad \|$

$\quad\quad\quad\quad qqq + rrr = -3qra + aaa$

The first substitution (on f. 103), $e = q$, $\dfrac{bb}{e} = r$ comes to the same thing as Problem 13 in the *Praxis* ($\dfrac{ee+bb}{e} = a$). The working, though different and dependent on the Canonical equation Harriot has generated, arrives at $eee = ccc + \sqrt{(ccccc - bbbbbb)}$, as does line 6 of p. 100 (Problem 13).
Harriot writes (in modern notation):

$$e^3 + b^6/e^3 = 2c^3 \qquad\qquad \text{And: } e^6 + b^6 = 2c^3 e^3$$

$$=> e^6 - 2c^3 e^3 + c^6 = -b^6 + c^6$$

$$=> e^6 - 2c^3 e^3 + c^6 = c^6 - b^6$$

$$1^{st} \; e^3 - c^3 = \sqrt{(c^6 - b^6)}$$

$$e^3 = c^3 + \sqrt{(c^6 - b^6)}$$

$$e = \sqrt[3]{(c^3 + \sqrt{(c^6 - b^6)})} = q$$

$$2^{nd} \; c^3 - e^3 = \sqrt{(c^6 - b^6)}$$

$$c^3 - \sqrt{(c^6 - b^6)} = e^3$$

$$e = \sqrt[3]{(c^3 - \sqrt{(c^6 - b^6)})}$$

Let $e = r$, $\dfrac{b^2}{e} = q$

$$=> b^6/e^3 + e^3 = 2c^3 \qquad \text{And } b^6 + e^6 = 2c^3 e^3$$

$$=> e^6 - 2c^3 e^3 = -b^6$$

As above.

$=>$ if first root $= q$

second $= r$ one $= q$, other $= r$

But $a = q + r$, etc.

$$a = \sqrt[3]{(c^3 + \sqrt{(c^6 - b^6)})} + \sqrt[3]{(c^3 - \sqrt{(c^6 - b^6)})}\ldots\ldots\ldots$$

No account is taken, as the *Praxis* does, of $c = b$, in which the Problem is followed by numerical examples, the first of which has $c > b$ and the other two have $c < b$. Note 2 (p. 101, *Praxis*) explains that these two exemplify what is called the irreducible case. The subsequent six numerical examples show how real roots emerge from conjugate complex numbers. All the numerical examples are the same in the *Praxis* as in the papers but in the latter, Harriot had probably taken the explanation for granted as there is no reference to it nor to "impossibility".

In the manuscripts, Add. MS 6783, ff. 104, 105, 108, 109 consider alternative methods for solution of the cubic. f. 104 carries out a straightforward substitution.

f. 105 brings back an equation in e^6 and provides a somewhat different approach. There is no indication as to why Harriot used yet another method for the same problem.

Add. MS 6783 f. 104 is e. 7 (top left) to be followed by e. 8 which is an apparently unrelated inequality.

f. 110–112 (e. 12, e. 13 and e. 14) have no equivalent in the *Praxis*. They refer to:

$$2.ccc = +3.bba - aaa.$$

19. This is the first of 21 reductions of biquadratics presented in such a form that a straightforward substitution will remove the term as required. All of these might be said to carry out the first step towards a solution of the biquadratic but none goes further than the removal of the term in a^3. And this despite the reference to Stevin's work on biquadratics referred to in the Preface.

20. Add. MS 6783, f. 113 (f. only top left) has

$$xxxz = +4.raaa + aaaa$$
$$xxxz = -4.raaa + aaaa$$

The first of these equations is an equation of the same type as those found in Problems 14 and 15 but the second has no equivalent in the book.

Examples are given:

$$9261 = +20, aaa + aaaa \qquad a = +7 - 21$$
$$9261 = -20, aaa + aaaa \qquad a = -7 + 21$$

This is one of many cases that imply that Harriot knew how to change positive into negative roots and vice versa. Equations like those above he calls conjugate. These particular equations were certainly constructed by assuming values for the two real roots.

f. 161 (f. only top left) has: $hhhk = +baaa + aaaa$
$$xxxz = +4.raaa + aaaa$$

which correspond to Problems 14 and 15, giving both general and particular solutions providing two roots. f. 145 has the same equations but is inconclusive.

21. Add. MS 6783 f. 114 (f. top left) has: $xxxz = -4.raaa - aaaa$
$$xxxz = 4.raaa - aaaa$$

The first corresponds to Problem 17 and the second to Problem 16. They both show numerical examples. In both the above, the manuscript page makes the "homogeneous term" positive whereas the *Praxis* has the highest power positive.

f. 119 (f. top left) has: $xxxz = +4.raaa - aaaa$ (Pr. 16)

$xxxz = -e.raaa - aaaa$ (Pr. 17)

As in f. 114, (f) the roots are raised and lowered and examples are given. The treatment of the homogeneous term and highest power is the same as in the previous problem.

Finally, corresponding to Problem 16, on f. 143 (f. top left) is found:

$$hhhk = +baaa - aaa$$
$$xxxx = +4.raaa - aaaa$$

Here the work is inconclusive.

22. Add. MS 6783, f. 115 (f. top left) has: $xxxx = +ccda + 4.raaa + aaaa$

$xxxx = -ccda - 4.raaa + aaaa$

which correspond to Problems 18 and 19 respectively. The same method is used and examples are given:

f. 141 (f. top left) has: $hhhk = +ccda + 4.raaa + aaaa$

$xxxx = +ccda + 4.raaa + aaaa$

Two roots are given, one positive and one negative. The page corresponds to Problem 18.

f. 148 (f. top left) has: $xxxz = -ccda + 4.raaa + aaaa$

$xxxz = +ccda - 4.raaa + aaaa$

$xxxz = +ccda - 4.raaa - aaaa$

$xxxz = -ccda + 4.raaa - aaaa$

These correspond to Problems 18, 19 and 33 and the work is unfinished.

23. Add. MS 6783, f. 116 (f. top left) has: $xxxz = +ppaa + 4raaa + aaaa$

$xxxz = +ppaa - 4raaa + aaaa$

These correspond to Problems 20 and 21 respectively but no examples are offered.

f. 117 (f. top left) has: $xxxz = +ppaa + 4raaa + aaaa$

$xxxz = +ppaa - 4raaa + aaaa$

Two pairs of conjugate equations are solved very elegantly, giving four roots in each case, positive and negative and changing positive into negative roots and vice versa. By now, it is clear that many of the biquadratics in Section 6 follow one another in "conjugate" pairs.

A pleasing feature of this page, in the second solution, is the use of zero in the modern manner.

f. 140 (f. top left) has: $hhhk = + cdaa + 4raaa + aaaa$

$xxxz = + cdaa + 4raaa + aaaa$

again corresponding to Problem 20 and two solutions are given, one positive and one negative.

24. Add. MS 6783, f. 118 (f. top left) has:

$$xxxz = + cdaa + ppaa + 4raaa + aaaa$$
$$xxxz = - cdaa + ppaa - 4raaa + aaaa$$

corresponding to Problems 22 and 23 and, again, though general work is done, no examples are given.

f. 130 (f. top left) has: $xxxz = + cdaa + ppaa + 4raaa + aaaa$
$$xxxz = - cdaa - ppaa - 4raaa - aaaa$$

corresponding to the same Problems, and without examples.

f. 139 (f. top left) has: $hhhk = + ffga + cdaa + 4raaa + aaaa$
$$xxxz = + ffga + cdaa + 4raaa + aaaa$$

corresponding to Problem 22 and provides an example in which only one root is given.

25. Add. MS 6783, f. 150 (f. top left) has:

$$hhhk = + ffga - cdaa + 4baaa - aaa$$
$$xxxz = + ffga - cdaa + 4raa - aaaa$$

corresponding to Problem 24. There is some general work.

ff. 151, 2 (f) give numerical examples of biquadratics, each with four real roots.

f. 153 (f) has the second of the above biquadratics again, corresponding to Problem 24 and gives numerical examples with four real roots.

ff. 154 (f) and 155 (f) give further numerical examples.

f. 157 (f) is an important page as Harriot writes about the "four-root rule" and refers to the fact that there are two "noetic" (i.e., in today's terms imaginary) roots.

26. Add. MS 6783, f. 200 (f) has: $xxxz = +ccda - ppaa + 4raaa + aaaa$
$$xxxz = -ccda - ppaa - 4raaa + aaaa$$
$$xxxz = -ccda + ppaa - 4raaa - aaaa$$
$$xxxz = +ccda + ppaa + 4raaa - aaaa$$

The first two lines correspond to Problems 25 and 30. The editor(s) seem not to grasp the possibility of changing the sign of the roots.

27. Add. MS 6783, f. 201 (f) has, corresponding to Problem 26,
$$hhhk = + ffga + cdaa - 4raaa + aaaa$$
f. 160 (f) has $xxxz = + ffga + cdaa - 4raaa + aaaa$

corresponding to Problem 27, and the work is unfinished.

f. 162 (f) has the second of these equations again, with some standard numerical solutions.

f. 201 (f) is unfinished.

There are other pages appearing less relevant.

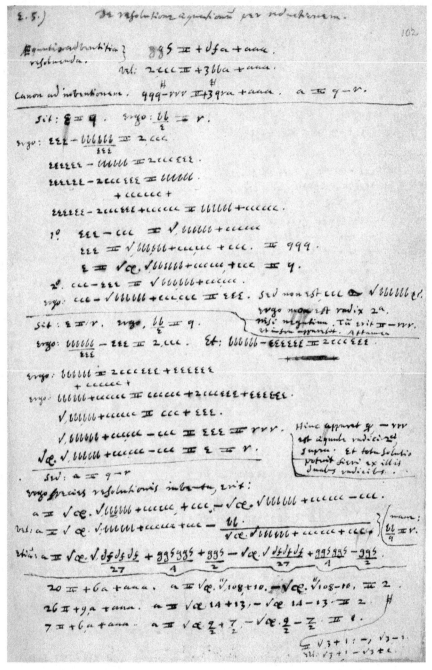

Figure 2. The British Library Board. All rights reserved. Add. Ms 6783, f. 102.

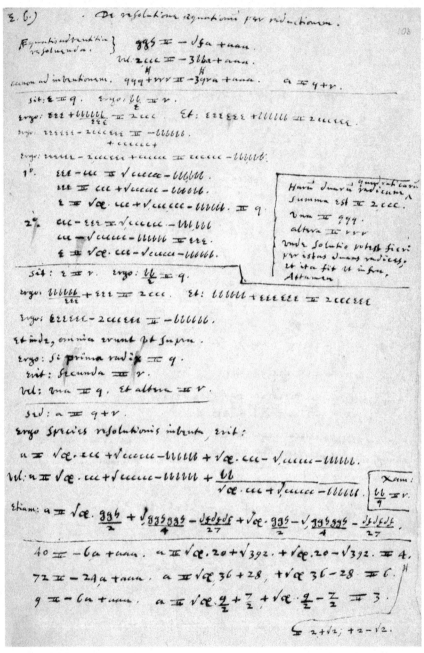

Figure 3. The British Library Board. All rights reserved. Add. Ms 6783, f. 103.

Notes on Numerical Exegesis

1. With reference to the use of the word "Exegesis," see Note to Definition 7. This word is never used by Harriot in the manuscript pages in the same way that it is used in the *Praxis*. (It is used algebraically in Add. MS 6784, ff. 19 and 23.) Harriot's manuscript pages are all headed by the title of Viète's work, *De numerosa potestatum resolutione*. Actually the title of Viète's 1600 publication was *De Numerosa Potestatum ad Exegesim Resolutione* and that of 1646 was *De Numerosa Potestatum Puravum atque Adfectarum ad Exegesim Resolutione Tractatus*. The translation given by Witmer (p. 311) is *On The Numerical Resolution of Powers by Exegetics.*

In the manuscripts Harriot refers all his examples to Viète and each page on which an equation is solved gives the number of the Problem in Viète's *De numerosa potestatum resolutione* that corresponds to Harriot's equation. Each page on which numerical equations are solved is headed by the title of Viète's work. Although virtually all the equations solved by Harriot in the manuscripts are identical to Viète's, this is not so for the equations solved in the *Praxis*, in which almost all the equations differ numerically from those found in Viète and, therefore, also from those in the manuscripts. Three problems in the *Praxis* solve exactly the same equation as in Viète and the manuscripts and those are a) Problem II, case of devolution (Viète, 1646, p. 179); Problem 7, case of devolution (*Praxis*, p. 136–137); and Add. MS 6782, f. 394 in the papers; b) Problem XI (Viète, 1646, p. 198); Problem 8 (*Praxis*, p. 143) and Petworth, f. 8/b.5; c) Problem XVI (Viète, 1646, pp. 211–212); Problem 4, (*Praxis*, p. 128); and Add. MS 6782, f. 46/c. 2. R.C.H. Tanner takes this as possible evidence that Warner had used Harriot's preliminary notes for the Numerical Exegesis, notes which have now been lost (Tanner Papers, University of Liverpool 599/51).

The most notable feature of this part of the *Praxis* is the presence of the polynomials at the head of each problem, algebraically derived from the equation whose "type" is expressed generally and which provides the basis for the solution of the particular equation. The terms of the polynomial then appear in column formation providing implicit, calculative instructions in an unambiguous algebraic symbolism.

There are 46 model polynomials for 46 different "types" of equation at the end of the book, covering 13 pages, for easy reference, perhaps. (All the same, the table does not include any polynomials or equations of the fifth degree.)

Harriot's manuscripts use the polynomials as an aid to or, rather, means of solution but without the subsidiary directions provided in the *Praxis*, such as indicating that a certain value represents a "tenfold single root," "subtraction," and so forth. Harriot, clearly, had no need to provide such subsidiary help for himself but the presence of these aids reinforces the notion of the *Praxis* as, in some sense at least, a teaching work, although perhaps for advanced students.

The manuscripts containing Harriot's work on the solution of equations with numerical coefficients are to be found in the British Library (Add. MS 6782 ff. 388–417) and there are more in the Public Record Office in Chichester, on loan from Petworth.

Ff. 388–399 (in the British Library) are numbered from 1 to 12 in the reverse order and refer to Viète's Problems 1 to 6. From f. 417 backwards there are labels in the top left-hand corner from c. 1 to c. 16 (this last is f. 402). But whereas we would expect f. 401 and f. 400 to correspond to Harriot's c. 17 and c. 18 respectively, these are in opposite order, so it is obvious that f. 400 and f. 401 have been bound in the wrong order.

Ff. 388–399 or, rather, ff. 399–388 have solutions of quadratic equations through to biquadratics almost entirely consisting of computations relating to solutions, and in which the work is preceded on the page by a check on the apparently known solution (known from Viète's work) together with the "Canonical Form". Such is not the case for the c. 1–c. 18 pages. (Such a check might be seen as numerical poristic as distinct from symbolic poristic in Section 4 of the *Praxis*.)

Ff. 400–417 (c. 1–c. 18) cover Viète's Problems 16–20 in the opposite order. It is only on these pages that there is any attempt to refer back to mathematics which would later form the content of the first part of the *Praxis*, as Secondary Canonical Equations are referred to. Work is done to establish limits between which the roots must lie. Nothing like this is given in the *Praxis*.

Manuscript papers in the Public Record Office (MF 1232), ff. 1–13 correspond to b. 12–b. 1 in the opposite order (and with an extra page between f. 11 and f. 12 and also two pages marked b. 3, one of which is largely numerical and looks like rough work. The comparative table below shows the equations solved in the *Praxis*, those solved by Viète and those in Harriot's manuscripts.

In the copy of these manuscript papers made by Torporley, described in the Introduction above (pp. xx), the numerical solutions are copied exactly from the Harriot manuscripts on pages 49v to 52r (of Torporley's *Summary*) and are all in order up to Problem VI. Just as in the Harriot manuscripts, there are no equations corresponding to Viète's Problems VII, VIII, and IX. Torporley begins "Prob. 10" on f. 50r and gives what we find in (Petworth) b. 1, etc., exactly as expected and then continues with the work found in papers now in the British Library.

We can put all the manuscript papers together in the right order on the assumption that Harriot worked through Viète from Problem I to Problem XX. And since Torporley gives a more-or-less exact copy (apart from his use of a Roman Superscript), which also omits Problems VII–IX, it is probable that all the manuscript papers now between the collections in the British Library and Petworth were then together.

If we assume that the Torporley papers show us Harriot's latest intentions, the Numerical Exegesis was intended to be at the end of the treatise Harriot planned. This conclusion is reinforced by the fact that some of the manuscript pages on Numerical Exegesis are labeled b and some c and some only with numbers, which indicates that labeling did not imply intention.

We may summarize as follows with respect to the contents of the *Praxis*, on Numerical Exegesis

a) Problems 1, 5, 11, 15 solve "pure" powers, using Viète's method but with different coefficients;
b) Problems 2, 7, 3, 8, 16, 4, 9 correspond to equations found in Viète and Harriot manuscripts, but with different numbers;
c) Problems 6, 10, 12, 13, 14 and the final equations with approximate solutions have no corresponding patterns in Viète or the Harriot manuscripts
d) Problem 7 (devolution) (in the *Praxis*) ⎤
   Add. MS 6782, f. 394.6 (Harriot)   solve the identical equation
   Problem II (Viète) ⎦
e) Problem 8 (anticipation) (*Praxis*) ⎤
   Petworth f. 8/b. 5 (Harriot)   solve the identical equation
   Problem XI (Viète) ⎦
f) Problem 4 (*Praxis*) ⎤
   Add. MS 6782, f. 416/c. 2 (Harriot)   solve the identical equation
   Problem XVI (Viète) ⎦

It would be easier to suggest reasons for Warner's alterations had they been consistent.

2. The solution to this "pure" equation is obtained digit-by-digit, that is, using a method of successive approximation. The explanation of the method is shown clearly in the symbolism.

The assumed root is initially separated into b plus c, b corresponding to the first digit in the root (taken from the left) and c the remainder. The largest value of b is found so that b^2 can be subtracted from 48233025 using b as the "trial divisor". The remainder after subtraction is 12233025. It remains to find the largest value of c so that $2bc + c^2$ can be subtracted from this using $2b$ as the "trial divisor," and since c refers to the second digit in the root, b has to be multiplied by 10 at this stage.

The process is continued until the remainder is zero, and for each digit the previous figures obtained have to be multiplied by 10 in order to relate them to the column under consideration.

The points over the original homogenous term indicate the placing of the digits in the solution. This is so because:

The square-roots of numbers up to, but not including 100 appear in the units column.
The square-roots of numbers up to, but not including 10,000 appear as tens.
The square-roots of numbers up to, but not including 1,000,000 appear as hundreds and so on.

Clearly, the largest b such that bb may be subtracted from the homogeneous term is 6. After subtraction we are left with 12233025. To find the value of c, an approximate divisor $2b$ is used (ignoring cc, as an approximation because c is relatively small), and taking b as 60 at this stage, gives us $c = 9$. Thus, Bc (taken as equal to $2bc + cc$) may be subtracted from the previous remainder and so on. The process continues, taking the new "B" as 690 and finding a new "c" until nothing remains.

No comparable case of an "unaffected" equation or "pure power" is to be found in the surviving Harriot manuscripts. The case shown in the *Praxis* differs numerically from that of Viète (Problem I) (who has $x^2 = 2916$, in modern notation) but the layout and procedure may be compared.

To compare the processes, we "translate" Viète's work into the notation of the *Praxis* and the exact correspondence of the procedures is clear:

$$aa = 2916$$
$$2916 = ff$$

Putting $b + c = a$

$$
\begin{array}{ll}
b + c & \\
b + c = 2916 & \\
\underline{+bb} & + 2.bc \\
Ab & +\ldots cc = 2916 \\
\hline
Bc &
\end{array}
$$

Root of the whole equation			5 4
Homogenous term to be solved	ff		2 9 1 6

Divisor	$b = 5$	
First individual root	$b = 5$	
To be subtracted	$bb = 25$	

Single root	5
Remainder of the homogenous term to be solved	4 1 6

Divisor	$2b$	1 0 0
Single tenfold root $b = 50$	$2.bc$	4 0 0
Second single root $c = 4$	cc	1 6
To be subtracted	bc	4 1 6

Root of whole equation finally extracted	5 4
The final remainder of the homogenous term	0 0 0 0

3. This is hardly a Lemma in the accepted sense of the word. It simply asserts that the solution may be verified by "backward composition" (the reverse of the way of composition, see Witmer, 1983, p. 316) in fact echoing Viète's words following the presentation of his model:

"Itaque si IQ, aequetur 2,916. fit IN 54, ex retrograda, quae omnino observata cernitur, compositionis via". (Thus, if $x^2 = 2916$, $x = 54$ as may be seen by the procedure of backward decomposition.)

4. *Note 1:* This note adds nothing to the text.

5. *Note 2:* attempts an explanation of the procedure for extracting the square root but falls far short of Viète's in clarity and attention to detail.

6. *Problem 2*: The procedure is a suitably modified form of the method used for finding the roots of "pure" powers. The first digit (from the left) in the solution is the largest value of b such that $db + bb$ can be subtracted from the homogene. A is used as the "trial" divisor for estimation. After subtraction, 3288208 is left.

Note the two rows of dots, the upper row marking the "square" term and the lower applicable to the linear term involving d (432 in this case). To obtain the first digit, 432 is placed so that its unit column is under the first (linear) dot. 3 ($b = 3$) is placed under the first in the upper row of dots, which corresponds to the square term. Such alignment must be observed throughout.

To obtain the second digit in the solution, d (432) is put in its proper place, the 2 under the second (from the left) linear dot, b is now 30 relative to this new column and we need the largest value of c so that $dc + 2bc + cc$ may be subtracted from 3288208. B, that is ($d + 2b$) is now the divisor for estimation and $c = 4$. Thus, the first two digits in the four-figure solution are 3, 4. Note that cc is ignored as an approximation in using B as "trial" divisor.

34 is now treated as if it were the first figure b in order to find the third digit. Clearly $b = 340$ in the third column and $c = 7$ enables the subtraction to be performed. The procedure is repeated until the remainder is zero. The stage is set for a general method based on the expansion of certain binomial powers.

Corresponding to this uncomplicated example are two manuscript pages, f. 399 and f. 398, both of which differ from it numerically. f. 399 presents the equivalent of $a^2 + 24a = 2356$ and precedes the two solutions (involving two alternative layouts) by a presentation of what would follow from taking the first digit in the solution as 4 instead of 3 (which is correct). Nothing is crossed out and the attempt is followed by a sentence, as if a text book were being written, to the effect that b must, therefore, be less than 4, that is 3. The page is completed by two alternative presentations of what is, in effect, the same method of solution. As is almost always the case, the A and B, used in the *Praxis*, is absent. Harriot clearly did not need this notational device. It is worth noting that Viète very often shows trials (and errors).

The work found on Harriot's manuscript page Add. MS 6782, f. 398 is identical to Viète's Problem I, equivalent to $a^2 + 7a = 60{,}750$ in modern notation.

7. This provides verification of the solution by "backward composition". Curiously, it is entitled "Lemma," which by rights means a minor Proposition used in an argument or proof. The verification (by substitution) that is presented here most closely comes under the definition of Poristic.

8. Cases can arise in which the coefficient of the linear term is very large and $db + bb$ will be too large to subtract from the homogenous term. As Viète expresses it: "This indicates that the square affects, rather than being affected, since it is less than the affecting plane" (Witmer, 1983, pp. 325–326).

In such a case, the first digit is "devolved" one decimal place to the right (or three, etc.) and the procedure is as before.

In Add. MS 6782, f. 397 Harriot presents a solution by devolution of the same type of equation but with different numerical values, $a^2 + 762a = 22{,}120$. The solution is carried out by devolution. At the side of the page, the corresponding Viète equation is given as: "$aa + 954.a = 18487$".

9. *Problem 3*: Nowhere is Harriot's superior notation more evident than in the equation presented in this problem, which involves negative signs. The working is clear as the sign '$-$' moves, as it does to this day, between being an operator and an indicator that a number is negative.

The corresponding (but not numerically identical) manuscript example is found in Petworth 13, where the equation is $aa - 7a = 60{,}750$, whose only difference from the *Praxis* lies in its omission of what Warner may have considered a help to the reader, the lines showing A and B, the exact or approximate trial divisors. The left-hand column tells the reader exactly what to calculate.

10. The procedure in this type of solution is very clearly explained. In the Problem to hand, three zeros are added to the left of the homogenous term and the first digit in the solution is 6.

None of Harriot's manuscript pages is identical to this numerically but Petworth 12 is similar: $a^2 - 240a = 484$. Harriot calls the method "Quadratum acephalum" (literally "headless quadratic".

Warner's explanation in the Praxis closely follows Viète's, except for its second sentence: "This indicates that the affecting plane is greater than the negatively affected square that is to be resolved [and it is then] rightly called an acephalic square" (Witmer, 1983, 341). Viète's own words are "acephalum quadratum" (acephalum is from ἀκέφαλος, without head or without leader).

Harriot / Warner's word "Anticipation" may have been considered a more user-friendly title.

11. Again, the explanation is clear and, although no justification is presented for the device to be used, it may well be that the rationale would run as follows:

$$a^2 - da = f^2$$
$$a^2 - da + d^2 = d^2 + f^2$$

But $d^2 - da$ is very small, so the root of $d^2 + f^2$ may be taken as an approximation. Witmer translates Viète (p. 343) as saying:

"… the coefficient may nevertheless sometimes extend beyond its place so that … in this case, the given negatively affected square must be added to the square of the linear coefficient…"

Harriot / Warner's words are very close to those of Viète: "artificium - - - epanorthicum" (p. 197) translated by Witmer (p. 343) as "corrective device" (Liddell and Scott give the derivation as "ἐπανορθωτις," setting right, correcting).

The *Praxis* does credit Viète with use of the term but, again, replaces it by another, which may have been considered more familiar to readers.

Viète's example following Problem X, $a^2 - 60a = 1600$ is repeated in Petworth 11, which works it out in full with some indication of what is being done. The top right hand corner of the page has (beneath 3): "Vieta exemplum ubi artificium parabola epanorthicum". (Viète's example where [he applies] a corrective technique.)

12. *Problem 4*: This is the first example in the *Praxis* which is replicated both in Harriot's manuscripts and in Viète's work [ff. 417–416 and Problem XVI (p. 211) in Viète, *Opera*, 1646].

The method in the *Praxis* is the same as before, taking account of the negative sign and the roots are obtained as would be expected. The usefulness of Harriot's symbolic notation is very evident here, and the solution uses the polynomials given by the binomial expansion.

There is no hint of a special case as Viète treats it in his Problem XVI, who heads the Section "Ad analysin potestatum avulsarum" ("avulsed powers", as translated by Witmer, 1983, p. 353) (Viète, *Opera*, 1646, p. 211). The word is not used in the *Praxis* or Harriot manuscripts. (Its literal meaning is "plucking off or tearing off, related to the branches of a tree," Lewis and Short.) The plucking or tearing off refers to the subtraction of the square from the linear term. No change of sign for the square is possible without the homogeneous term becoming negative. (The homogenous term is positive in the *Praxis*, MS and Viète.)

The corresponding pages in the Harriot manuscripts are ff. 416–417 (where the reading should be in the reverse order). Ff. 400–417 use the results from Section 4 (of the *Praxis*) and found in Add. MS 6783 ff. 163, to obtain the positive roots of the equation and at the top of these pages are statements of the Canonical equation and the positive roots. Ff. 388–399 and the Petworth pages omit this and have instead a sort of "poristic" demonstration that the root does, in fact, satisfy the equation.

The top of f. 417 has:

$$bc = ba \qquad a = b$$
$$+ca - aa \quad a = c$$

For if, $a = b$ we will have:
$$bc = bb$$
$$+bc - bb \text{ and it is so}$$

If $a = c$, we will have:
$$bc = bc$$
$$+cc - cc \text{ and it is.}$$

So, $a = b$ or $a = c$.

None of this appears in the *Praxis*, which appears not to use such earlier results except in the form of Lemmata. (See notes 3, 7, 19, 23, and 30).

13. *A Short Cut*: The reference to Theorem 2, Sect. 5 is mystifying as there is no apparent connection between anything in Section 5 and the equation

$-aa + da = ff$. A clear connection is drawn, though, between the roots, coefficient and the homogenous term of the particular quadratic equation under consideration.

Viète states implicitly, *but only in a particular case*, the result concerning what would later be called symmetric functions of the roots: [*Opera Mathematica*, 1646, p. 218] (Witmer, 1983 p. 360).

Quoniam 12 triplum quadratum e triente 6, majus est coefficiente plano 11: ideo IN de tribus lateribus potest explicare, quorum summa 6, trinum sub iis rectangulum 11 solidum sub iisdem factum continue 6.

14. *Problem 5*: The solution of this "pure" cubic is performed by expanding $(b + c)^3$. The approximation for the divisor to obtain c is $3.bb$, the terms involving c (i.e., $3bcc + ccc$) being omitted because c is small.

No corresponding material exists in the surviving manuscripts even though Viète performs the same process in Problem II of the first section of the *De numerosa* but with different numbers, i.e., $x^3 = 157,464$.

15. *Problem 6*: The calculation is performed as expected, again omitting terms involving c in the approximation. Neither the MS nor Viète's work contain an equation corresponding to the one solved here.

16. *Lemma*: The reference to "backward composition" echoes Viète's ubiquitous "Ex retrograda quae omnino observata cernitur compositionis via" (which may be perceived by everyone by the method/road of backward composition). The same procedure is followed in ff. 388–399 and Petworth 1–13 but in each case at the start of the proceedings.

17. For an explanation of the different method that is used here, see Note on p. 136 (wrongly printed as p. 134 in the original edition).

18. *Note*: In the first method for solving Problem 6 (see note 15), the first divisor is approximated to in a "heterogenous" manner but the second, alternative method (see note 17), although more difficult, preserves the law of homogeneity. (The first divisor in the first method is $ff + d + bb$ and in the second is $ff + db + bb$).

19. *Problem 7*: The corresponding example in the manuscripts is Add. MS 6782, f. 395: $a^3 + 30a = 14356197$ which is solved by Viète in Problem II (Viète, *Opera*, pp. 166–168).

20. In the original edition of the *Praxis* p. 138 has been wrongly numbered as p. 136. The work is a further example of verification "poristically."

21. *Example of Devolution*: This is a straightforward example, which is the second of three equations that are replicated numerically in both manuscripts (Add. MS 6783, f. 394) and by Viète in his Problem II.

22. *Rectification*: In the original edition of the *Praxis* p. 139 has been wrongly printed as p. 137. The corresponding example in the papers is in Petworth 7: $a^3 + 64a = 1024$, ("Vieta exemplum epanorthicum"), which is given in the *De numerosa* following Problem XI not in full tabular form but only as part of a discussion of *epanorthosis* (Vièta, *Opera*, 1646, p. 200).

23. *Problem 8*: The Harriot manuscript, Petworth 9, is of the same type as this Problem.

$a^3 - 10a = 13584$ as is Viète's Problem XI (Vièta, *Opera*, 1646, p. 198).

24. *Lemma*: A further example of poristic.

25. A Note to the effect that negative and positive terms are to be respectively "gathered together" for ease of calculation.

26. *Anticipation*: Harriot's manuscript page Petworth 8 ("Vieta exemplum cubus acephalus") and Viète both solve:

$$a^3 - 116,620a = 352947$$

(This is one of three equations that appear in identical numerical terms in the *Praxis*, manuscripts and Viète.)

27. *Rectification*: Harriot's manuscript page Petworth 7 ("Vieta exemplum ubi artifictum parabola epanorthicum") and Viète (Problem XI) solve:

$$a^3 - 6,400a = 153,000$$

28. *Problem 9*: The structural correlate in the Harriot manuscripts is:

$155,520 = 13,104a - aaa$, (Add. MS 6783, ff. 412–415).

Viète's Problem XVII solves the equivalent of:

$13,104x - x^3 = 155,520$.

29. *Short Cut*: The reference back to Section 4, Proposition 6, giving $ff = bb + bc + cc$ foreshadows later knowledge of the Symmetric Functions of the Roots (of an equation).

30. *Problem 10*: It is not clear why another straightforward case, corresponding to the type already demonstrated in Problem 8, should be solved at this stage. It can only be attributed to a sort of carelessness, coming as it does after the "avulsed" power case, which should, logically, terminate the list of cubics.

31. Another poristic Lemma.

32. *Problem 11*: The first "pure" biquadratic, corresponding to Viète's Problem IIII: $x^4 = 331,776$ (in modern notation). No corresponding example has come to light in the MS.

33. *Problem 12*: No structural correlate is to be found either in Harriot's papers or Viète's *De Numerosa*.

34. *Anticipation*: No structural correlate in the manuscripts or in Viète's *De Numerosa*.

35. *Problem 13*: No structural correlate is to be found for this in the work of Harriot or Viète. Note the extra approximation in the working.

36. *Problem 14*: No structural correlate to be found in Harriot manuscripts or Viète's work.

37. *Problem 15*: The corresponding Viète equation is equivalent to:

$$x^5 = 7,962,624 \text{ (Problem III, Viète, } Opera, 1646, \text{ pp. } 169\text{–}171).$$

38. *Problem 16*: Warner has chosen an equation whose structural correlate is found in Harriot's manuscript Petworth 1 and (identically) in Viète, Problem XV.

The work on Petworth 1, involving an "affected power" "gathers up" terms and labels them B, C and D but not in relation to any Canonical polynomial but as an aid to calculation and this is done throughout the Petworth papers.

39. *Approximations*: No structural correlates to be found.

Comparative Table of Equations Solved

A comparative table of polynomial equations with numerical coefficients as solved by Viète, by Harriot in his manuscripts and as they appear in the *Praxis*.

Viète Problem	Harriot MS Problem	*Praxis* Problem	Comment
Problema I (3r,v)(165 to 166)* 1Q, aequari 2916	None	Problem 1 (p. 117) $aa === 48233025$	
Problema II (3v to 4v) (166 to 168) 1C, aequari 157, 464	None	Problem 5 (p. 131) $aaa ==== 10568963 6352$	
Problema III (4v to 5r)(168 to 169) 1QQ, aequari 331, 776	None	Problem 11 (p. 151) $aaaa ==== 1956529 5376$	
Problema IV (5r to 6r) (169 to 171) 1QC, aequari, 7,962,624	None	Problem 15 (p. 160) $aaaaa ==== 1575550 9298176$	
Problema V (6r to 6v)(171 to 172) 1CC, aequari 101, 102, 976	None	None	

* Page references to the 1600 and 1646 editions of Viète's *De numerosa potestatum resolutione*.

Above are "pure" equations. Below are "affected" equations.

Unless indicated as "Petworth" all manuscript references are to MS 6782

(Continued)

Viète Problem	Harriot MS Problem	Praxis Problem	Comment
a) Problema I (7v to 9r) (174 to 175) $1Q + 7N$, adaequari 60,750	a) f. 398[2] $aa + 7.a = 60{,}750$	a) Problem 2 (p. 119) $aa + 432.a ===== 13584208$	Viète b) devolution MS b) f. 399 done before f. 398
b) (8r to 9r) (175 to 176) $954N + 1Q$, aequatur 18487	b) f. 399[1] $aa + 24.a = 2{,}356$	b) (p. 121) $aa + 75325.a ==== 41501984$	MS c) devolution (b): f. 399 done before f. 398 *Praxis* b) devolution c) devolution
	c) f. 397[3] $aa + 762.a = 22120$	c) (p. 122) $aaa + 675325.a ==== 369701983$	
a) Problema II (9r to 10r) (176 to 178) $1C + 30N$, aequari 14,356,197	a) f. 396[4.] $aaa + 35.a === 22932$	a) Problem 7 (p. 136) $aaa + 45796.a === 449324752$	Viète b) devolution MS b) the same as Viète a)
b) (10r to10v) (178 to 179) $95{,}400N + 1C$ aequantur 1, 819, 459	b) f. 395[5] $aaa + 30.a === 14356197$	b) (p. 138) $aaa + 95400.a ===== 1819459$	*Praxis* b) the same as MS c) and Viète b)
	c) f. 394[6] $aaa + 95{,}400a === 1819459$	c) (p. 139) $aaa + 274576.a === 301163392$	
a) Problema III (10v to 11v) (180 to 182) $1C + 30Q$, aequari 86,220,288	a) f. 393[7] $aaa + 30aa = 86220288$	None	Viète b) the same as MS b)
b) (11v to 12v) (182 to 183) $10{,}000Q + 1C$, aequatur 5,773,814	b) f. 392[8] $aaa + 10{,}000aa = 5773824$		
a) Problema IV (12v to 13r) (183 to 184) $1QQ + 1{,}000 N$ aequari 355,776	a) f. 391[9] $aaaa + 1000a === 355, 776$	None	Viète a) the same as MS b) Viète b) the same as MS b)

b) (13r to 14r) (184 to 185) 10000N + 1QQ aequentur 2731776	b) f. 390[10] $aaaa + 100{,}000a === 2731776$		
a) Problema V (14r to 14v) (186 to 187) 1QQ + 10C, aequetur 470,016	a) f. <u>389</u>[11] $aaaa + 10.aaa === 470{,}016$ b) f. <u>389</u>[11] $aaaa + 300aaa === 4{,}478{,}976$	None	Viète a) the same as MS a)
a) Problema VI (14v to 15v) (187 to 188) 1QQ + 200Q, aequetur 446,976	a) f. 388[12] $aaaa + 200, aa + 100.a ==$ 449,376	None	Viète b) the same as MS a)
b) (15r to 15v) (188 to 189) 1QQ + 200Q + 100N aequetur 449376			
a) Problema VII (16r to 16v) (190 to 191) 1QC + 500N, aequetur 254,832	None	None	
a) Problema VIII (16v to17v) (191 to 193) 1QC + 5C, aequetur 257,472	None	None	
a) Problema IX (17v to 18v) (193 to 195) 1CC + 6000N, aequetur 191,246,976			

(Continued)

Viète Problem	Harriot MS Problem	Praxis Problem	Comment
a) Problema X (18v to 19r) (195 to 196) $1Q - 7N$, aequari 60,750 ($x^2 - 7x = 60{,}750$)	a) Petworth ((PRO) f. 13/b. 1) $aa - 7.a ==== 60750$	a) Problem 3 (p. 124) $aa - 624.a ==== 16305156$	Viète a) the same as MS a) Viète b) the same as MS b) acephalum Viète c) the same as MS c) epanorthicum
b) (19v to 20r) (196 to 197) $1Q - 240N$, aequetur 484	b) Petworth ((PRO) f. 12/b. 2) $aa - 240a === 484$	b) (p. 125) $aa - 6253.a ==== 6254$	*Praxis* b) Anticipation c) Rectification
c) (20r) (197) $1Q - 60N$, aequari 1600	c) Petworth ((PRO) f. 11/b. 3) $aa - 60.a === 1600$	c) (p. 127) $aa - 732.a ==== 86005$	
a) Problema XI (20r to 21r) (198 to 199) $1C - 10N$, aequari 13,584	a) Petworth ((PRO) f. 9/b. 4) $aaa - 10.a === 13{,}584$	a) Problem 8 (p. 141) $aaa - 2648.a ==== 91148512$	Viète a) the same as MS a) Viète b) the same as MS b) Viète c) the same as MS c) MS b) acephalum c) epanorthicum
b) (21r to 22r) (199 to 200) $1C - 116{,}620N$, aequetur 352.947	b) Petworth (PRO) f. 8/b. 5 $aaa - 116{,}620a === 352{,}947$	b) (p. 143) $aaa - 116{,}620a = 352947$	[Petworth (f. 10/b. 3) is numerical only]
c) (22r) (200 to 201) $1C - 6400N$ aequari 153,000	c) Petworth (PRO) f. 7/b. 6 $aaa - 640a === 153{,}000$	c) (p. 145) $aaa - 127296.a ==== 85700000$	Praxis b) the same as Viète b) and MS b)

a) Problema XII (22r to 22v) (201 to 202) 1C − 7Q, aequari 14,580	a) Petworth (PRO) f. 6/b. 7 $aaa − 7.aa ==== 14580$	None	Viète a) the same as MS a) Viète b) the same as MS b) Viète c) the same as MS c) MS b) acephalum MS c) epanorthicum
b) (22v to 23v) (202 to 203) 1C − 10Q, aequari 288	b) Petworth (PRO) f. 5/b. 8 $aaa − 10aa === 288$		
c) (23v to 24r) (203 to 204) 1C − 7Q, aequetur 720	c) Petworth ((PRO) f. 4/b. 9) $aaa − 7.aa ==== 720$		
d) (24r) (204) 1C + 8Q aequari 1024	d)Petworth (PRO) f. 4/b. 9 $aaa + 8aa = 1024$		
a) Problema XIII (24r to 25r) (205 to 206) 1QC − 68C + 202, 752N, aequari 5,308,416	a) Petworth ((PRO) f. 3/b. 10) $aaaa − 68aaa + 202752a$ $= 5,308,416$	None	Viète a) the same as MS a)
a) Problema XIIII (25r to 26r) (207 to 208) 1QQ + 10C − 200N, aequari 1,369,856	a) Petworth ((PRO) f. 2/b. 11) $aaaa + 10aaa − 200a$ $= 1,369,856$	None	Viète a) the same as MS a)

(Continued)

Viète Problem	Harriot MS Problem	Praxis Problem	Comment
a) Problema XV (26r to 27r) (208 to 210) $1QC - 5C + 500N$, aequetur, 7,905,504	a) Petworth ((PRO)) f. 1/b. 12) $aaaaa - 5.aaa + 500.a = 7,905,504$	a) Problem 16 (p. 162) $aaaaa - 57.aaa + 5263.a ==== 900050558322$	Viète a) same as in MS a)
a) Problema XVI (27v to 28r) (211 to 212) 370N – 1Q. aequari 9261	a) (f. 417/c. 1) Algebra on limits of roots in Viète problem	a) Problem 4 (p. 128) $-aa + 370a === 9261$	From now on Harriot puts homogeneous term on left-hand side. Praxis puts highest power first, even if negative.
b) (28r, 28v) (212–214) 370N – 1Q. aequari 9261	b) (f. 416/c. 2) $9261 = 370a - aa$		Same equation throughout. Case of "avulsed" equation, that is, highest power 'torn' from other.
a) Problema XVII (29r to 29 v) (214 to 215) 13104N – 1C, aequari 155,520	a) (f. 415/c. 3) Algebra on limits of roots in Viète equation	a) Problem 9 (p. 146) $-aaa + 52416.a ===$ 1244160	MS a), b), c) refer to Viète equation. Another "avulsed" equation
b) (29v, 30r) (215–216) 13104N – 1C, aequari 155,520	b) (f. 414/c. 4) (more on f. 415/c. 3) (ready for f. 413/c. 5)		
	c) (f. 413/c. 5) $155.520 ==== 13104a - aaa$		
	d) (f. 412/c. 6) Same equation as before. Roots obtained in opposite order.		

a) Problema XVIII (30r to 30v) (216 to 217) 57Q – 1C, aequari 24,300	a) f. 411/[c. 7] Algebra on limits of roots	None
b) (30v, 31r) (217–218) 57Q – 1C, aequari 24,300 (alternatively)	b) f. 410[c. 8] Algebra on limits of roots	
	c) f. 409[c. 9] Algebra in relation to Viète	
	d) f. 408[c. 10] $2430057 = aa - aaa$	
a) Problema XIX (31v to 32r) (219) 27, 755N – 1QQ, aequari 217,944	a) f. 407[c. 11] Algebra on limits of roots	None
b) (32r to 32v) (220) 27, 755N – 1QQ, aequari 217,944	b) f. 406[c. 12] Algebra on limits of roots (ready for f. 405/c. 13)	
	c) f. 405[c. 13] $217,944 = 27755a - aaaa$	
a) Problema XX (32v to 33v) (221–2) 65C – 1QQ, aequari 1,481,544	a) (f. 404/c. 14) Algebra on b) (f. 403/c. 15) limits of roots c) (f. 402/c. 16) algebra re. Prob XX	None
b) (33v – 34r) (222–3) 65C – 1QQ, aequari 1,481,544	d) (f. 400/c. 17) $1481544 = 65aaa - aaaa$	ff. 400 and 401 are out of order.
	e) (f. 401/c. 18) $1481544 = 65aaa - aaaa$	

Textual Emendations

(C means present in original Errata correctly.
*C means present in original Errate incorrectly.)

	Correct	Incorrect	
1) Pref. p. 1, 1.17	illum in certamine	illum certamine	
2) Pref. p. 2, 1.25	permansit dum	permansit. Dum	
3) p. 1, 1.–3	constituta	constitua	
4) p. 2, 1.16	tanquam	tantum	C
5) p. 2, 1.–14	dato scil.	datum scil.	C
6) p. 4, 1.–3	$-faaa$	$+faaa$	C
7) p. 5, 1.4	reductitis. Varij.	reductitia varij	
8) p. 5, 1.–7	in quinta	in quarta	C
9) p. 5, 1.–4	Aequatio	AEquatio	
10) p. 13, 1.–14	$a - r$	$a - a$	
11) p. 13, 1.–1	$bcdf$	$bedf$	
12) p. 17, 1.4	designata, posito	designata. Posito	
13) p. 17, 1.20			
14) p. 17, 1.23	$+bca$	$-bca$	
15) pp. 18–19	$+bca$	$-bca$	
(five times)	$+bda$	$-bda$	C
	$+cda$	$-cda$	
16) p. 19, 1.–8	$-bcd$	$-bda$	
17) p. 20, 1.4	posito	Posito	
18) p. 20, 1.8	, posito	. posito	
19) p. 21, 1.–1	Ut	U	
20) p. 25, 1.11	$bdaa$	$bfaa$	
21) p. 25, 1.–8	[Delete $+cda$]	$+cda$	
22) p. 27, 1.7	5. 9. 15 & 18	5. 9. & 10	*C
23) p. 29, 1.7	[Correct as	ba corrected	
	it stands]	to aa	*C
24) p. 30, 1.22	$-bbcc$	$-bccc$	
25) p. 32, 1.2	PROBLEMA	PROBEMA	
26) p. 32, 1.14	$aaa + baa$	$+baa$	*C
27) p. 32, 1.–4	binomia	trinomia	*C
	(no error)	(alleged)	
28) p. 36, 1.6	$-bfaa$	$+bfaa$	C
29) p. 39, 1.8	$cdfa$	$cdfd$	
30) p. 41, 1.10	$cdfa$	$cafa$	
31) p. 47, 1.–12, –11	$(-cdaaa + bbcda$	$-cdaaa + bbcda$	
	$(b + c + d \, b + c + d$	$b + c + d \, b + c + d$	
32) p. 48, 1.–1	$b + c - d$	$b + c + d$	
33) p. 49, 1.18	$bd + cd - bc$	$bd + cd \, bc$	
34) p. 50, 1.–2($\frac{1}{2}$)	$bbb + bbc + bcc + ccc$	$bbb + bbc + bcc + ccc$	
35) p. 52, 1.8	$+bc$	$+bb$	

	Correct		Incorrect		
36) p. 53, 1.21	$+d$	$+d$	$-d$	$-d$	C
37) p. 56, 1.7	$=== +bcd$		Omitted		
38) p. 56, 1.–2	adhuc		adhnc		
39) p. 60, 1.22	$b - c$		$c + b$		
40) p. 63, 1.11	cdaa		caaa		
41) p. 64, 1.7	$b ==== f$		$c ==== f$		
42) p. 64, 1.23	bdfb		bdbb		
43) p. 68, 1.14–15	_____|		_____|		
44) p. 68, 1.18	si		Si		
45) p. 72, 1.14($^1/_2$)	_____		_____		
46) p. 72, 1.21	per 21		per 22		C
47) p. 74, 1.–17	$+bbccd$		$-bbccd$		
48) p. 74, 1.–11	enunciatum		enuntiatum		
49) p. 74, 1.–2	$bd + cd - bc$		$dc + cd + bc$		
50) p. 75, 1.11	$bd + cd - bc$		$bd + cd\ bc$		
51) p. 75, 1.–16	$b.$ vel $c.$ radici		$b.$ vel dici		C
52) p. 77, 1.–5	$a.$ in $c.$ erit		$a.$ in erit		C
53) p. 77, 1.–1	enunciatum		ennnciatum		
54) p. 78, 1.1	78		72		
55) p. 78, 1.–11	pp. maxima est, qq.		pq. Maxima est, cc.		
56) p. 80, 1.11	Prop. 15		Prop. 14		
57) p. 80, 1.2–4	$\begin{array}{l} rq \\ rq \\ rq \end{array}\Big\|$		$\begin{array}{l} rq \\ rq \\ rq \end{array}\Big\|$		
58) p. 80, 1.13	enunciatum		enunciarum		
59) p. 81, 1.6–7	$\dfrac{qqr + qrr \mid}{\;\;\;qqr + qrr \mid}{4}$		$\dfrac{r + qqr \mid}{\;\;\;r + qrr \mid}{4}$		C
60) p. 81, 1.23	$\dfrac{\begin{array}{l} qq + qr + rr \mid \\ qq + qr + rr \mid \\ qq + qr + rr \mid \end{array}}{27}$		$\dfrac{\begin{array}{l} qq + qr + rr \mid \\ qq + qr + rr \mid \\ qq + qr + rr \mid \\ qq + qr + rr \mid \end{array}}{27}$		)))C
61) p. 82, 1.–11	aequatio		Aequatio		
62) p. 84, 1.–18	$\dfrac{pq + pr + qr}{3}$		$pq + pr + qr$		C
63) p. 84, 1.15	igitur		igiitur		
64) p. 85, 1.1	85		Omitted		
65) p. 86, 1.–14	$-bccc$		$-bcc$		
66) p. 88, 1.7	aa		aaa		
67) p. 88, 1.13	a		aa		
68) p. 89, 1.7	$a === e + b$		$a = c + b$		
69) p. 89, 1.–11	$-ccc$		$-cce$		
70) p. 91, 1.–18	bb		$bb-$		
71) p. 91, 1.–3	$+$		$++$		
72) p. 92, 1.9	$a === b - e$		$a === b - c$		
73) p. 94, 1.5	transpositis,		transpositis.		
74) p. 94, 1.12	$ee +$		$ee + +$		
75) p. 94, 1.16	reliquis,		reliquis.		
76) p. 95, 1.2	reliquis,		reliquis.		
77) p. 95, 1.8	secundo		secundo,		

	Correct	Incorrect	
78) p. 95, 1.13	transpositis,	transpositis.	
79) p. 95, 1.–11	$-e - b$	$+e - b$	
80) p. 97, 1.2	$-eee$	eee	
81) p. 97, 1.13	$aaa + 3baa\ldots$	$aaa + baa\ldots$	
82) p. 97, 1.15	dde	ddd	
83) p. 97, 1.–3	$eee - 3.bee$	$eee - bee$	
84) p. 98, 1.3,4	$eee - 3.bbe$ $+..dde$	$aaa - 3.bba$ $+..dda$	
85) p. 99, 1.4	$\sqrt{3}) - ccc + \sqrt{ccccc} +$ $bbbbbb$	$\sqrt{ccc} + \sqrt{} - ccccc +$ $bbbbbb$ (also other cube root sign missing)	C
86) p. 99, 1.5	$\sqrt{3})\, ccc + \sqrt{ccccc} +$ $bbbbbb$	$\sqrt{ccc} + \sqrt{ccccc} +$ $bbbbbb$	
87) p. 99, 1.6	$\sqrt{3}) - ccc + \sqrt{ccccc} +$ $bbbbbb$	$\sqrt{ccc} + \sqrt{-ccccc} +$ $bbbbbb$	C
88) p. 99, 1.7	$\sqrt{3}) - ccc + \sqrt{ccccc} +$ $bbbbbb$	$\sqrt{ccc} + \sqrt{ccccc} +$ $bbbbbb$ (also other cube root sign missing)	C
89) p. 99, 1.13	$\dfrac{\sqrt{81}}{4}$	$\dfrac{\sqrt{3}.\,81}{4}$	
90) p. 99, 1.–4	$-3.bb$	$+3.bb$	
91) p. 99, 1.–4	$-3.bba$	$+3.bba$	
92) p. 99, 1.–11 to p. 100, 1,14	e looks like c]		
93) p. 100, 1.15	minor	maior	
94) p. 100, 1.–8, –9, –10,–11	$\sqrt{3}.)$	Cube root sign omitted 6 times	
95) p. 100, 1.–2	72	27	
96) p. 101, 1.–7	392	292	
97) p. 102, 1.3	$\sqrt{5408}$	5408	
98) p. 102, 1.13	$-3.bbbb$	$+3.bbbb$	
99) p. 102, 1.–7	$-4.baaa$	$+4.baaa$	
100) p. 103, 1.–9	$+3.bbe$	$+bbe$	
101) p. 104, 1.15	$+aaaa$	$+4.aaaa$	
102) p. 104, 1.16	$+12.bbbe - ..4bbbb$	$-12.bbbe + ..bbbb$	
103) p. 104, 1.–14	requisita	requisita.	
104) p. 104, 1.–3	$+4beee - 12.bbee +$ $12.bbbe$	$+4beeee - 12.bbee -$ $12.bbbe$	
105) p. 105, 1.7	imperata	imparata	
106) p. 106, 1.11	$+bb$	$+be - bb.$	
107) p. 107, 1.11	$-bbff$	$-3.bbff$	
108) p. 107, 1.16	$6.bbee$	$6.bee$	
109) p. 108, 1.–6	$-3.bbe$	$-bbe$	
110) p. 108, 1.–5	$+aaaa$	$+4.aaaa$	
111) p. 111, 1.12	imperata	Imparata	
112) p. 112, 1.11	$-ddda$	$-ddaa$	
113) p. 112, 1.–11	$2.\mathit{ffbe}$	ffbe	
114) p. 114, 1.7	$-4.\mathit{ffbb}$	$-\mathit{ffbb}$	
115) p. 114, 1.–18	$e === +b - a$	$e == a + b$	

	Correct	Incorrect	
116) p. 114, 1.–17	imperata	imparata	
117) p. 115, 1.15	*aa*	*a*	
118) p. 115, 1.21	*eeee*	*eee*	
119) p. 116, 1.11	*−..dddb*	*−..dddd*	
120) p. 117, 1.7	educere	reducere	C
121) p. 118, 1.3	12233025	42233025	
122) p. 118, 1.4–7	Columns correct	Columns incorrect	

Correct	Incorrect
12233025	42233025
120	120
1080	1080
81	81
1161	1161

	Correct	Incorrect	
123) p. 118, 1.12	16	81	
124) p. 120, 1.4	432	0432	
125) p. 120, 1.–17–18	Columns	Columns	
126) p. 120, 1.–16	7232	7332	
127) p. 121, 1.9, 10, 13, 14	⌋ ⌋ ⌋	∟ ∟ |	
128) p. 121, 1.–$\frac{1}{2}$	....	.....	
129) p. 122, 1.5	5	3	
130) p. 122, 1.6	75825	75325	
131) p. 122, 1.14	76325	76328	
132) p. 122, 1.21	75325	75323	
133) p. 122, 1.24	527275	597277	
134) p. 122, 1.25	7560 (+Column)	4760 (Column)	
135) p. 123, 1.19	2701300	2701400	
136) p. 123, 1.29	$c === 7$	$c === 6$	
137) p. 124, 1.–7	7376	7379	
138) p. 124, 1.–1	498356	498456	
139) p. 124, 1.–$\frac{1}{2}$	....		
140) p. 125, 1.4–10	Columns	Columns	
141) p. 125, 1.8	$2.bc$	bc	
142) p. 126, 1.4	Columns	Columns	
143) p. 126, 1.6	A −253	A 253	
144) p. 126, 1.9	$−Ab$ −1518	Ab 1518	C
145) p. 126, 1.12	Column	Column	
146) p. 126, 1.13	60	50	
147) p. 126, 1.15	−12506	−57506	
148) p. 126, 1.18	$−Bc...$ −11894	$Bc...$ 11894	C
149) p. 126, 1.18	Column	Column	
150) p. 126, 1.19	62	54	
151) p. 126, 1.20	334854	334864	
152) p. 126, 1.26	Column	Column	
153) p. 127, 1.–$\frac{1}{2}$	...		
154) p. 128, 1.7	−2196	−2199	
155) p. 128, 1.12	4665	3665	
156) p. 128, 1.12$\frac{1}{2}$	————		
157) p. 128, 1.17	$c === 5$	$b === 5$	C
158) p. 128, 1.18	25	35	

	Correct	Incorrect	
159) p. 128, 1.23	$-aa + da$	$aa + da$	C
160) p. 128, 1.–$^1/_2$	..		
161) p. 129, 1.5		b === 2	
162) p. 129, 1.12	370	371	
163) p. 129, 1.16	$c === 7$	$c - - - 7$	
164) p. 129, 1.16	$-2bc$	$2bc$	
165) p. 129, 1.–4	(Delete)	b === 3	
166) p. 129, 1.–1	-11739	11739	C
167) p. 130, 1.12	-939	939	C
168) p. 131, 1.–5	105689636352	10568963636352	
169) p. 132, 1.10	$3.bbc$	$3.bb$	
170) p. 132, 1.11	$3.bcc$	bcc	
171) p. 132, 1.–12	534681600	534681630	
172) p. 132, 1.–11	906240	996240	
173) p. 133, 1.4	$b + c$	$b + 6$	
174) p. 133, 1.14	186394079	186334079	
175) p. 133, 1.–$^1/_2$	. . . 42218079	. . . 42218079	
176) p. 135, 1.10–13	Columns	Columns	
177) p. 135, 1.14	21760	27600	
178) p. 135, 1.16	1250	1255	
179) p. 135, 1.17	1441760	1446760	
180) p. 135, 1.19	Columns	Columns	
181) p. 135, 1.21	decupl.	dupl..	
182) p. 135, 1.–$^1/_2$	6751199	6751199	
183) p. 136, 1.1	136 (page no.)	134	
184) p. 136, 1.8	30464	36464	
185) p. 136, 1.–7	analytice	analyce	
186) p. 137, 1.–$^1/_2$	. . . 10211712 . . .	. . . 10211712	
187) p. 138, 1.1	138 (page)	136	
188) p. 138, 1.7	1690816	1690818	
189) p. 138, 1.–10, –11, –15, –16, –17	⌐ ⌐ ⌐	L L |	
190) p. 138, 1.–7	congrua	cougrua	
191) p. 139, 1.1	139 (page)	137	
192) p. 139, 1.9	95500	95400	
193) p. 139, 1.23		8	
194) p. 140, 1.16–18	Columns	Columns	C
195) p. 140, 1.–14	3211428	3226428	
196) p. 141, 1.10–12	⌐ ⌐	L L	
197) p. 141, 1.–7	bb	$3.bb$	
198) p. 141, 1.–5	$-ffb$	ffb	
199) p. 142, 1.–1,–2,–3	⌐ ⌐	L L	
200) p. 143, 1.–6	$-116620a$	$- 116620$	

	Correct	Incorrect	
201) p. 145, 1.6	85760000	85700000	
202) p. 145, 1.12	45118016	45444672	
203) p. 145, 1.−16	*Ab*	*−Ab*	
204) p. 145, 1.−4	23877	1377	
205) p. 145, 1.−1	4349880	4349888	
206) p. 146, 1.3	4349880	4349888	
207) p. 146, 1.8	716994	716694	
208) p. 146, 1.9	−763776	−763676	
209) p. 146, 1.14	4349880	4249888	
210) p. 146, 1.15	universalis	uuiversalis	
211) p. 146, 1.−10 to −12	⌐ ⌐	L L	
212) p. 147, 1.12	*−bb*	*−3.bb*	
213) p. 147, 1.15	*b === 2*	*b === 4*	
214) p. 147, 1.21	*−3b*	*3b*	
215) p. 147, 1.−12	*b === 210*	*b === 450*	
216) p. 147, 1.−6	*c === 6*	*c === 2*	
217) p. 147, 1.−4	−816696	−816096	
218) p. 147, 1.−2	2 1 6	2 4	
219) p. 148, 1.10	104032	104832	C
220) p. 148, 1.20½	[Omit lines]	———	C
221) p. 148, 1.21	*−ccc*	*−cc*	
222) p. 149, 1.8−10	⌐ ⌐	L L	
223) p. 150, 1.14	1350	1450	
224) p. 150, 1.15	*ccc*	*cc*	
225) p. 150, 1.−11	−432480	−432400	
226) p. 150, 1.−6	434928	434848	
227) p. 151, 1.6,7,8,12, 13	⌐ ⌐ ⌐	L L |	
228) p. 152, 1.−6	*cccc*	*ccc*	
229) p. 153, 1.8−11	⌐ ⌐	L L	
230) p. 153, 1.13	2068948	206848	
231) p. 153, 1.−9	*hhhh*	*−ggg*	
232) p. 153, 1.−4	*b === 3*	*b === 5*	
233) p. 153, 1.−1, −2	3· ·	3· ·	
	1 2 7 1 7 2 8	1 2 7 1 7 2 8	
234) p. 154, 1.−12	*4.bbbc*	*4.bbbb*	
235) p. 154, 1.−6	*−A*	*−B*	
236) p. 154, 1.−4	*bbbb*	*b bb*	
237) p. 155, 1.3½	4 9 8 4 8 7 8 8 0 8	4 9 8 4 8 7 8 8 0 8	C
238) p. 155, 1.−16	128634056	128534046	
239) p. 155, 1.−10	345951216	345851216	
240) p. 156, 1.6−9	⌐ ⌐ ⌐	L L L	
241) p. 157, 1.20	1064204778	1064204776	
242) p. 157, 1.31	203441134	203441138	
243) p. 157, 1.33	202682350	202682354	
244) p. 157, 1.35	25600	2560	
245) p. 158, 1.12−15	⌐ ⌐ ⌐	L L L	

	Correct	Incorrect
246) p. 159, 1.7	−108654	108654
247) p. 159, 1.25	−43778	−47778
248) p. 159, 1.27	−430080	−430000
249) p. 159, 1.−4	4846338	4850338
250) p. 160, 1.6	−757760	−257760
251) p. 160, 1.10	203434880	203441138
252) p. 160, 1.11	765038	758784
253) p. 160, 1.12	202669842	202319042
254) p. 162, 1.4	Columns	Columns
255) p. 162, 1.5	Columns	Column
256) p. 162, 1.7	2150	1480
257) p. 162, 1.8	1719 - - - -	1717 - - - -
258) p. 162, 1.12	5.$bcccc$	4.$bcccc$
259) p. 162, 1.−8 to −12	⌋ ⌋ ⌋	∟ ∟ ∟
260) p. 162, 1.−11	$b + c$	$b + b$
261) p. 163, 1.3	$-3.ffbbc$	$+3.ffbbc$
262) p. 163, 1.3	900050558322	90050558322
263) p. 163, 1.18	160000 - - - -	260000 - - -
264) p. 163, 1.22	$bbbbb$	$bbbbc$
265) p. 163, 1.28	- - - - - - 57	- - - - 1 - - - - 57
266) p. 163, 1.−14	21052	11052
267) p. 163, 1.−10	3200000	3720000
268) p. 163, 1.−½	· · · 1 0 4 5 7 4 8 6 3 2 0 2 · · ·	· · · 1 0 4 5 7 4 8 6 3 2 0 2
269) p. 164, 1.3½	· · · 1 0 4 5 7 4 8 6 3 2 0 2 · · · · · ·	· · · 1 0 4 5 7 4 8 6 3 2 0 2 · · · · · ·
270) p. 164, 1.10	576000	976000
271) p. 164, 1.12	1672762 - - -	1672772 - - -
272) p. 164, 1.19	$c === 6$	$b === 6$ C
273) p. 164, 1.−7	in biquadratico	in quadratico C
274) p. 165, 1.−7	B	A
275) p. 165, 1.−3	17861	17866
276) p. 166, 1.6	Omit	1
277) p. 167, 1.7	5055	505
278) p. 167, 1.10	$c === 5$	$b === 5$
279) p. 167, 1.12	27900	27800
280) p. 167, 1.14	14540000	15540000
281) p. 167, 1.20	$c === 2$	$b === 2$
282) p. 167, 1.23	Bc	B
283) p. 167, 1.27	$b === 4520$	$c === 4520$
284) p. 169 (No.5) 1.2–3	[Surprising approx]	
285) p. 169 (No.6), 1.7	$-2.dbc$	$2.dbc$
286) p. 169 (No.8), 1.4	bb	db
287) p. 169 (No.9), 1.5	$-ffb$	ffb
288) p. 169 (No.9), 1.7	$-ffc$	$-ff$
289) p. 170 (12), 1.5	$-ffb$	$-ff$
290) p. 171 (5), 1.4	$ffbb$	$gggb$

	Correct	Incorrect
291) p. 171 (5), 1.11	4.*bccc*	4.*bc c*
292) p. 171 (7), 1.4	*ffbb*	*ff*
293) p. 172 (13), 1.4	−2.*ffb*	−2*fb*
294) p. 172 (14), 1.11	*Bc*	*B*
295) p. 173 (15), 1.9	*ffcc*	*ff c*
296) p. 173 (17), 1.4	2.*ffb*	2.*f b*
297) p. 173 (19), 1.10	*dccc*	Omitted
298) p. 174 (21), 1.9	*gggc*	*ggggc*
299) p. 174 (21), 1.10	−*dccc*	−*d cc*
300) p. 174 (23), 1.4	*bbb*	−*bbb*
301) p. 174 (23), 1.15	4.*bccc*	4.*b cc*
302) p. 175 (25), 1.1	*ffaa*	*ggga*
303) p. 175 (27), 1.6	−*dbbb*	−*d bb*
304) p. 176 (29), 1.18	−*cccc*	*cccc*
305) p. 176 (31), 1.11	−2*ffbc*	−2*ffb*
306) p. 176 (32), 1.15	3.*dbcc*	2.*dbcc*
307) p. 177 (36), 1.11	−*gggc*	.*gggc*
308) p. 178 (37), 1.20	*cccc*	*cc c*
309) p. 178 (39), 1.19	4.*bccc*	4.*b cc*
310) p. 179 (41), 1.18	6.*bbcc*	6.*b cc*
311) p. 179 (43), 1.2	*ggg*	*gggg*
312) p. 179 (43), 1.13	−2.*ffbc*	−2.*ffb*
313) p. 180 (46), 1.14	*dccc*	*d cc*
314) p. 180 (46), 1.21	−*cccc*	−*ccc*

Appendix

Table X shows the contents of the British Library manuscripts Add. MS 6782, 3 and 4, which contain almost all of Harriot's work in algebra and also the Harriot papers on solution of equations which were separated from the others in the British Library and copies are now in the Public Record Office at Chichester. The left-hand column gives the folio number in the manuscript, the next column gives the number in the top left-hand corner of the manuscript page as given by Harriot, and the third gives a brief account of the content. Not every page is included, as the selection has been on the basis of the existence of such a page-numbering by Harriot, the relevance of the content to the contents of the *Praxis*, or both of these. (Virtually all of the verso pages are scrap or "waste" as Harriot spoke of in his Will.)

Inevitably, such a selection process has a subjective element, for instance, in estimating what is scrap or waste. There are also some unknowables such as whether some papers have been destroyed. We can do no more than our best in this situation. We have to take what there is as evidence, order the material as best we can in accordance with such evidence and trust that our judgement of what is irrelevant or waste is not too far from the truth.

TABLE X

Add. MS 6782

Folio	Top left-hand label (by Harriot)	Content
2	f c. 11	Numerical calculations
[3–6	-	" "]
7	f c. 1	Numerical calculations
8	f c. 2	" "
9	f c. 5	" "
10	f c. 6	" "

Folio	Top left-hand label (by Harriot)	Content
11	f c. 9	" "
12	c. 1	Calculations (numerical)
13	c. 1	"
14	c. 2	"
15	c. 3	"
16	c. 4	"
17	c. 4) 2D	"
18	c. 5	"
19	c. 6	"
20	c. 6) 2	"
21	c. 7	"
22	f b. 1	Lists of numbers (?) meaning unclear
23	f b. 2	Calculations (numerical)
24	f b. 1	"
25	f b. 5	"
26	f a.1) f b. 1	"
27–106	-	Letter and number – squares
84–88	1) to 4)	Combinatorial?
		Miscellaneous calculation. Waste pages?
		Combinations and Transpositions
		Figurative numbers, progressions
107–144	1–36	Triangular numbers and progressions, some not in Harriot's hand?
		Miscellaneous algebra (not Theory of Equations)
147	A.1	" "
148	A.1	" "
149, 150	A2, A3	" "
181–183	-	Multiplication of binomial factors up to $(b + c + d)^4$
187	c. 3.d	Miscellaneous equalities (waste?) [Curious sign?]
188	c. 2	Algebra (waste?)
189	c. 2.aliter	Algebra (waste?)
190	c. 2	Algebra? (waste?)
191	c. 3	Algebra? (waste?)
192	c. 4	" "
264	Euclid:lib:13	Geometrical work
314	B	Canonical equations? Waste?
315	C	Algebra
316	D	" "
317	E	" "
318	F	" "
319	G	" "
320–324	aa − ae	" "
[325–359]	-	Miscellaneous material on Combination, Transposition, quantities in continued proportion, etc., in Harriot's hand?
360	f. 8	Solution of a biquadratic
361	-	Lists of cubics and quadratics. Waste?
362–375	-	Work on Infinity (in Harriot's hand?)
376–378	A2, A3, A4	Miscellaneous algebra. Waste?

379	A4	" "
380	-	2 egs. of "Zetetic". Geometrical
388–399	12 to 1	*De numerosa potestatum resolutione*
400–417	c. 17, c. 18, c. 16 to c. 1)	" "
[418–505]	-	Miscellaneous algebraic geometric and trigonometric work. Reference to Zetetic. Much waste.

Add. MS 6783

f. 1, f. 2	b. 1, b. 2	Headed De Radicalibus; addition and subtraction of surds (literal and numerical)
f. 3	b. 3	*De Additione et Subtractione* (Continuation of previous pages)
f. 4	b. 4	Multiplicatio
f. 5	b. 5	Headed *Applicatio* Division of square and fourth roots
f. 6	b. 6	Headings *Multiplicatio* and *Applicatio* (mid-page). Multiplication and division of sums and products of square and fourth roots of numbers. (waste?)
f. 7	b. 7	Similar to above
f. 8	b. 8	*Radicem extractio* (Geometrical method)
ff. 9–11	b. 9–11	The same
f. 12	$4)^{20}$	Algebra. Waste?
f. 22	2	Algebra. Waste?
f. 34	-	Solution of cubic following Bombelli
ff. 49–60	-	Lists of equations, linear to biquadratic with solutions
f. 65	e	*De Resolutione aequationem per reductionem* ($ccd = +baa - aaa$)
f. 67	e	Same as f. 65 (but $ccd = +baa + aaa$)
ff. 72–75	-	Cubics with numerical coefficients f. 72r has six equations corresponding to those on pp. 101–102 of *Praxis*
f. 94	4.	*De Resolutione per reductionem.* Biquadratic, see beginning Section 6.
f. 96	3.	As f. 94 (solving biquadratic)
f. 97	2.	As above (solving biquadratic)
[ff. 98–112	e. 1 to e. 14 (2 x e. 7)	Heading: *De resolutione equationem per reductionem*]
f. 98	e. 1	Lists of +, − (combinatorially?) and lists of numbers. Also, as heading: "Apparatus, ad genera, species et differentiis aequationum adventitiarum"
f. 99	e. 2	Lists of literal equations, linear to biquadratic
f. 100	e. 3	Lists of literal equations, linear to quintic
f. 101–112	e. 4 to e. 14	Work on solution of cubics. f. 106, f. 107, e. 8 and e. 9 have inequalities

ff. 113–119	f	Same heading. Work on biquadratics (with solutions)
ff. 130–162	f	" "
ff. 163–183	d. 21 to d. 1	*De Generatione aequationum canonicorum*
ff. 184–185	e. 29, e. 28	Inequalities (see Section 5) Reduction (heading)
ff. 186–198	e. 27 to e. 15	Reduction of cubics. Algebraic and numerical examples
ff. 199–202	f	Reduction of biquadratic. Algebraic
f. 204	d. 7.2°	[Relevant to f. 174]. Reduction of biquadratic
f. 215	e. 3.2°)	Heading: *Exempla aequatiunum in numeris*
f. 216	e. 3.3°)	Lists of numerical equations with solutions, cubics give 1 or 3 solutions. and biquadratic 2 or 4 solutions. Negative solutions given
f. 217	e. 3.4°	" "
f. 218	e. 3.5°	" "
f. 219	e. 3.6°	" "
ff. 220–228	A, B, C, 3, 2) 1), 1), 1), 1)	Harriot's hand? Word "Exegesis" used
ff. 233–234	a, b	Geometrical (to do with quadratics)
f. 235	c	Work on quadratics, biquadratics and higher degree (up to 8th)
ff. 236–237	d, e	Geometrical
f. 268	f. 8 (middle & top-right)	Biquadratics with 1 root given in each case
f. 272	D	Algebra. Waste?
ff. 273–275	a^2r, ar, a^3r	To do with removal of terms from biquadratics. Waste?
f. 280	-	(Upside-down). Considerable written work on Viète's condition for roots of cubic
f. 281	4	Waste? Harriot's hand?
ff. 282–296	as to ah	Waste? Removal of terms from equations up to biquadratic
ff. 302–304	bb^2 (all three)	Waste?
f. 305	bb^2	List of equations from cubic to sixth power
f. 307	bb^2	Waste?

Add. MS 6784

| ff. 322–325 | 1.) 2.) 3.) 4.) | Approximates to Section 1 of *Praxis* |

Petworth/Public Record Office

| ff. 1–13 | b. 12 … b. 4, b. 3)2, b. 3, b. 2, b. 1 | Solution of polynomial equations with numerical coefficients |

TABLE Y

The following table presents the Torporley version of Harriot's algebra (the *Summary*) in which Harriot's page-numbering is repeated. The left-hand column gives the folio number in Torporley's papers, the middle column gives (where possible) the Harriot page-numbering, repeated by Torporley.
Torporley MS Arc/L.40.2/L/40

Folio	Top left-hand	Contents
35r	?	Heading: *Operationes logisticae in notis*. Corresponds to Section 1. Also corresponds to MS pages
35	b. 3	Continues the above. One-third of way down page, new heading: *De Radicalibus*. Poem on multiplying positive and negative signs.
36r	b. 3 (?)	Work on 4 rules with surds
(lower down) 36r	b. 4	Multiplication of surds
(lower down still) 36r	b. 5	Division of surds
(further down) 36r	b. 1? b. 6?	More on surds
36v	b. 7, b. 8, b. 9	Work on surds
(under double line across page) 36v	1), 2)	" "
37r	?	" "
37v	6), 7), 8), 9)	" "
38r	?	" "
38v	)))) (no number)	" "
39r	?	" "
39v	?	" "
40r	?	" "
40v (down left-hand page)	6, 7 (?), 8, 9, 10} 11, 12, 13, 14, 15}	" "
41r	?	" "
(down 41v)	d1, d. 2, d. 3, d. 4	*De generatione aequatiorum canonicorum*
42r	d. 5, d. 6, d. 7	Further on above
42v	d. 10, d. 11, d. 12	" "
43r	d. 13 to d. 19	" "
43v	d. 20, d. 21, then d. 7.2	Further on above

Folio	Top left-hand	Contents
44r	?	(Cannot see top left-hand)
	e. 1	Reminiscent of 6783, f. 98, e. 1
	e. 2	
	e. 3	Seems like combinatorial work on signs and numbers
	e. 4	On Generation of Canonicals. Echoes 6783, f. 101
	e. 5 (?)	Like 6783, f. 102 but with Roman index
	e.. 6 (?)	notation (e''' for eee). (*Praxis*, p. 99, Problem 12)
44v	e.. 6 (?)	Appears on f. 103.
	e.. 7	Appears in f. 104} copied exactly
	e.. 7.2	Appears in f. 105} apart from notation
	e.. 8	Proves Lemmata but not inequality
	e.. 9	The same as f. 107 (but for notation)
	e.. 10	Comparison of adventitious with Canonical equation (?) No mention of equipollence (f. 108).
45r	e. 11	(f. 109)
	e. 12 – 1.14	Study of cubics
45v	e.. 15	" "
	e. 16	(Not in Praxis)
	e. 17	Study of cubics
	e. 18	" "
	e. 19	" "
46r	?	" "
	?	" "
46v	e. 27	" "
ff. 46v	e. 27	" "
	e. 28	6783, f. 184 & f. 185 copied exactly. NB
	e. 29	Section 5 work not separated from section 6 work
	f. 1	f. 113, biquadratics
	f. 2d to f. 5	" "
f. 47r	?	" "
f. 47v	f. 8v to f. 15r	Biquadratics. Following f. 13r comes d. 13.2)
f. 48r	?	Biquadratics
f. 48v	(Near to/f.le to f. 4e)	Biquadratics
f. 49r	?	Biquadratics, *ec*.
f. 49v	1) to 12)	Up to Pr. 6 (Viète). (Up to f. 388) (Problems VII, VIII, IX of Viète missing as in MS and *Praxis*.)
f. 50r	"pr. 10" up to "pr. 13"	Gives Petworth b. 1, etc. then back to BL MS. Cossic and Roman superscripts.

f. 50v	b.11	Problems 14 and 15
	b. 12	Then c. 1 to c. 6, problems 16 and 17
f. 51r	?	Problems 18, 19, 20
f. 51v	c. 16, 17, 18	Problem 20. Work on (non-numerical) equations, quadratic to biquadratic.
f. 52r	?	More algebra
52v	mid-page AB)	Geometrical
53r	?	Algebra and geometry
53v	6) to 14)	Contents need further study
54r	?	Combinations
54v	?	Combinations and Transpositions followed by summary of previous contents (Headings)

Table Z is a summary of Harriot's (top left-hand) page-numbering and the corresponding content of his papers (as in Add. MS 6782, 3 and 4 in the British Library).

To summarise:

1. Some page-numbering of Harriot's refers to several topics, i.e., Items 2, 4, 5, 9, 11, and 14 (Table Z).
2. The material in Section 1 of the *Praxis* is numbered as in Item 4 (Table Z).
3. The material in Sections 2–4 of the *Praxis* is numbered as in Item 13 (Table Z).
4. Some of the material in Section 5 of the *Praxis* is numbered as in Item 11. [Notably, Add. MS 6783 f. 104 (e. 7) is followed by f. 105 (e. 8) on inequalities.]

TABLE Z

Page numbering	Contents	Relevance
1. f.c. number	Numerical calculations	
2. c. number	Numerical calculations/inequalities/numerical solutions of equations	Some relevant
3. f.b. number	Numbers and calculations	
4. Number alone	Rules in algebra/triangular numbers and progressions/numerical equations/algebra in general/work on biquadratics	Some relevant
5. A. number	Triangular numbers and progressions. Miscellaneous algebra and geometry	
6. B, C, D	Canonical equations? Waste?	
7. *aa* to *ae, as* to *ah*	" " "	
8. f. number	Biquadratics	Relevant
9. b. number	De Radicalibus. Solution of numerical equations	Relevant
10. e	Cubics/Geometry	Relevant
11. e. number	Cubics/Inequalities/Lists of equations. Biquadratics (6783, ff. 215–219)	Relevant
12. f	Biquadratics	Relevant
13. d. number	Generation of Canonicals	Relevant
14. a, b, c, etc.	Geometry/Lists of quadratics and upwards	
15. *ar*, etc.	Removal of terms from biquadratics	Relevant
16. *bb*	Algebraic waste	

5. The material in Section 6 (*Praxis*) is numbered as in either Items 8), 10), 11), 12) or 15) of Table Z. (Thus, several topics may have the same type of page-numbering, as in 1 above or one topic may have several types of page-numbering as here.)

6. Finally, the solution of numeral equations may be numbered as in Items 2), 4) or 9).

It may certainly be inferred, then, that the top left-hand page-numbering in the manuscript pages does not *necessarily* imply Harriot's intentions as to order of presentation in a possible future publication (as argued by J. Stedall, 2003), but may indicate a form of numbering corresponding to his immediate writing.

It is unclear why the Numerical Exegesis in the *Praxis*, whose page-numberings in the manuscripts are b's and c's and numbers alone, should come after the folios with the page-numberings e's and f's. Also, although almost all biquadratic equations are treated under the heading f, a few (Add. MS 6783, ff. 215–19) are included under e. The explanation might very well lie in the fact that those letters are something other than simple indications of Harriot's intended order. Also, if the letters represented Harriot's intentions, why would some numerical equations have b, others have c and a third batch have numbers alone? The order (as may be seen from the second column of the table of equations, pp. 262–7 is "number alone" then "b. number" and then "c. number"). Perhaps Harriot numbered the pages at the top-left hand corner after doing the work, but changed his mind about the order. This is not the numbering, in the compilation used by Torporley. The large number of questions that we are asking rather than answering here is indicative of the range of possibilities that appears to be consistent with the facts as we presently know them.

Taking the Torporley *Summary* as a guide, the order of topics envisaged by Torporley would probably be:

1. Operationes logisticae in notis (Operations of arithmetic in signs).

2. De radicalibus (On Roots).

3. Generatione aequationum canonicorum (Generation of canonical equations) (which would include the solutions of the cubic equations now in Section 6; and some comparison of common with canonical equations involving inequalities as in Section 5).

4. More on cubic equations.

5. Biquadratic equations.

6. Numerical exegesis (i.e., a series of worked examples).

What is certain is that Warner (the putative editor of the *Praxis*) tampered with Harriot's intended order. He may have extracted from the same source as Torporley or from the manuscripts as they have come down to us, or from a different source entirely, which is not available to us.

In any event, Warner made the following changes to the manuscript material:

1. He added a Preface and Definitions.

2. He cut down the material for Section 1.

3. He omitted *De radicalibus*.
4. He altered and stretched out into three Sections (now 2, 3, and 4) what was more concisely done in the manuscripts.
5. He added the idea of "equipollence" in Section 5.
6. He put together in Section 6 what were intended to be separate sections on cubic and biquadratic equations. He cut this work on the reduction of such equations to ones more easily solved (except for Problems 12 and 13) by changing the root and omitting much work on cubics and biquadratics.
7. In the Numerical Exegesis, he added equations (corresponding to "pure powers") not considered by Harriot; altered the order of presentation from one in which equations with positive coefficients were all done first and followed by those with negative coefficients (following Viète); added others of a pattern in neither Viète nor Harriot and omitted others; and in almost all cases altered the numbers involved.

One may deplore Warner's tampering with Harriot's material and his contravention of Harriot's intentions. However, the intrinsically probabilistic character of our reasoning forces us to a merely provisional condemnation. Such are the conclusions which may be drawn concerning Harriot's intentions for his future treatise and Warner's failure to carry them out in the *Praxis*. It is all the more fortunate, then, that Torporley made his own suggestions in the *Corrector Analiticus* concerning the contents of the work that Warner should have produced. In 2003, Dr. Jacqueline Stedall published *Harriot's Treatise on Equations*, which reproduces the Harriot manuscripts which conform to Torporley's suggestions and which is as close as we are ever likely to get to Harriot's intentions.

The following table connects pages, Propositions and Problems in the *Praxis* with the corresponding manuscript pages when they are judged to have a significant connection.

Praxis	MS	Comment
Section 1		
pp. 7–8	Add. MS 6784, f. 322	a) Add. MS 6783, ff. 1–11, also correspond to Section 1 passage in
p. 8–9	f. 323	Torporley's *Summary*. (Left-hand
p. 10	f. 324	side of page b. 1 to b. 11).
p. 11	f. 325	b) In 6784, left-hand side of page gives 1, 2, 3, 4.
Section 2		
p. 12	Add. MS 6783, f. 183 (d. 1), f. 182 (d. 2), f. 181 (d. 3), f. 180 (d. 4), f. 178 (d. 6)	
p. 13	f. 178 (d. 6), f. 177 (d. 7)	
p. 14	f. 176 (d. 8), f. 175 (d. 9), f. 171 (d. 13)	
pp. 14–15	f. 156, 13.2°	MS pages omit both cases giving negative and imaginary roots cf. f. 301

p. 16 (Pr.1)	f. 183 (d. 1)	
pp. 16–17 (Pr.2)	f. 183 (d. 1)	
p. 17 (Pr.3)	f. 180 (d. 4)	
pp. 17/18 (Pr.4)	f. 181 (d. 3)	
pp. 18–19 (Pr.5)	f. 182 (d. 2)	
p. 19 (Pr.6)	f. 178 (d. 6)	
pp. 19–20 (Pr.7)	f. 178 (d. 6)	
p. 20 (Pr.8)	f. 178 (d. 6)	
pp. 20–21 (Pr.9)	f. 176 (d. 8)	
p. 22–23 (Pr.10)	f. 175 (d. 9)	
p. 23–24 (Pr. 11)	f. 177 (d. 7)	
p. 24–25 (Pr.12)	f. 177 (d. 7)	
p. 25 (Pr.13)	f. 171 (d. 13)	
p. 26 (Pr.14)	f. 171 (d. 13)	
p. 26 (Pr.15)	f. 171 (d. 13)	
p. 27 (Three special Equations)	f. 101	f. 101 uses reduced form of the first of the special equations

Section 3

p. 29 (Pr.1)	f. 183 (d. 13)	
pp. 29–30 (Pr. 2)	f. 181 (d. 3)	
pp. 30–31 (Pr.3)	f. 181 (d. 3)	
p. 31 (Pr.4)	f. 180 (d. 4)	
p. 32 (Pr.5)	f. 179 (d. 5)	
p. 33–4 (Pr.6–8)	f. 101 (e. 4)	f. 101 used for Problem 12, Section 6, solution of cubic
p. 34–35 (Pr.9)	f. 177 (d. 7)	
pp. 36–37 (Pr.10)	f. 177 (d. 7)	
pp. 37–38 (Pr.11)	f. 177 (d. 7)	
p. 39 (Pr.12)	f. 176 (d. 8)	
pp. 40–41 (Pr.13)	f. 176 (d. 8)	
pp. 41–42 (Pr.14)	f. 176 (d. 8)	
pp. 42 (Pr.15)	f. 175 (d. 9)	
p. 43 (Pr.16)	f. 175 (d. 9)	
p. 44 (Pr.17)	f. 175 (d. 9)	
p. 45 (Pr.19)	f. 174 (d. 8)	
pp. 45–46 (Pr.20)	f. 173 (d. 7)	f. 204, d. 7.2° alternative
p. 46 (Pr.21)	f. 172 (d. 6)	
pp. 46–51 (Lists)	f. 170 (d. 14), f. 168 (d. 16), f. 169 (d. 15)	

Section 4

p. 52 (Pr.1)	f. 183 (d. 1)	
p. 52 (Lemma)	f. 183 (d. 1)	
pp. 52–53 (Pr.2)	f. 183 (d. 1)	
p. 53 (Lemma)	f. 183 (d. 1)	
pp. 53–54 (Pr.3)	f. 180 (d. 4)	
p. 54 (Lemma)	f. 180 (d. 4)	
pp. 54–55 (Pr. 4)	f. 181 (d. 3)	
p. 55 (Lemma)	f. 181 (d. 3)	
pp. 55–56 (Pr.5)	f. 182 (d. 2)	
pp. 56–57 (Lemma)	f. 182 (d. 2)	

p. 62 (Pr.18)	f. 178 (d. 6)	
p. 62 (Pr.19)	f. 178 (d. 6)	
pp. 62–63 (Pr.20)	f. 178 (d. 6)	
p. 63 (Pr.21)	f. 176 (d. 8)	
pp. 63–64 (Lemma)	f. 176 (d. 8)	
pp. 64–65 (Pr.22)	f. 177 (d. 7)	
p. 65 (Lemma)	f. 177 (d. 7)	
pp. 65–66 (Pr.23)	f. 175 (d. 9)	
pp. 66 (Lemma)	f. 175 (d. 9)	
p. 67 (Pr.24)	f. 177 (d. 7)	
pp. 67–68 (Lemma)	f. 177 (d. 7)	
pp. 68–69 (Pr.25)	f. 176 (d. 8)	
pp. 71–72 (Pr.29)	f. 176 (d. 8)	
p. 72 (Pr.30)	f. 176 (d. 8)	
pp. 72–73 (Pr.31)	d. 175 (d. 9)	
pp. 73–74 (Pr.32)	f. 175 (d. 9)	
p. 74 (Pr.33)	f. 175 (d. 9)	
pp. 76–77 (Pr.38)	f. 171 (d. 13)	
p.77 (Pr.39)	f. 171 (d. 13)	
p. 77 (Pr.40)	f. 171 (d. 13)	

Section 5

p. 79 (Lemma 2)	f. 106r, (e. 8)	All four sheets containing the
p. 79 (Lemma 3)	f. 106r, v (e. 8)	inequalities are headed: "De
p. 80 (Lemma 4)	f. 106r (e. 8)	resolutione equationum
pp. 81–82 (Lemma 5)	f. 107r (e. 9)	reductionem". See *Summary*,
pp. 84–85 (Lemma 6)	f. 184r (e. 29)	ff. 44v–46v. also Birch MS 4394,
pp. 85–86 (Lemma 7)	f. 184r (e. 29)	f. 392

Section 6

p. 87	f. 163	
p. 88	f. 164	
p. 89 (Pr.1)	f. 91, f. 198 (e. 15)	cf. Viète, *Opera*, 1646, p. 130, II
p. 90 (Pr.2)	f. 67 (e), f. 197 (e. 16)	" " p. 130, I
p. 91 (Pr.3)	f. 65 (e), f. 196 (e. 17)	p. 130, III
pp. 91–92 (Pr.4)	f. 195 (e. 18)	" " p. 130, I
pp. 92–93 (Pr.5)	f. 194 (e. 19)	" " p. 131, VII
p. 93 (Pr.6)	f. 193 (e. 20)	" " p. 131, II
pp. 93–94 (Pr.7)	f. 192 (e. 21)	" " p. 131, VI
pp. 94–95 (Pr.8)	f. 191 (e. 22)	" " p. 131, IV
pp. 95–96 (Pr.9)	f. 190 (e. 23)	" " p. 131, V
		Negative root
		implied
pp. 96–97 (Pr.10)	f. 189 (e. 24)	" " p. 131, III
	f. 187	Zero used as a
		calculable entity in f. 187
pp. 97–98 (Pr.11)	f. 188 (e. 25)	Negative root implied in *Praxis*
pp. 98–99 (Pr.12)	f. 102 (e. 5)	
	f. 411	

pp. 99–102 (Pr.13)	f. 103 (e. 6)	
	f. 104 (e. 7)	
	f. 105 (e. 7), f. 108 (e. 10)	
	f. 109 (e. 11)	
	f. 402	
pp. 101–102	ff. 72–73.	
	See also f. 74, f. 75 r	
	and v	
pp. 102–103	f. 113 (f)	
(Pr.14,15)	f. 161 (f), f. 145 (f)	
pp. 103–4 (Pr.16)	f. 119 (f), f. 143 (f)	
p. 104 (Pr.17)	f. 114 (f)	
pp. 104–105 (Pr.18)	f. 115 (f)	
	f. 141 (f)	
pp. 105–106 (Pr.19)	f. 148 (f)	Uses±
p. 106 (Pr.20)	f. 116 (f), f. 117 (f),	
	f. 140 (f)	
pp. 106–107 (Pr.21)	f. 148	
pp. 107–108 (Pr.22)	f. 118 (f), f. 130 (f)	
	f. 139 (f)	
pp. 108–109 (Pr.23)	f. 118 (f)	
p. 109 (Pr.24)	f. 150 (f), f. 151 (f),	
	f. 152 (f), f. 153 (f),	
	f. 154 (f), f. 155 (f),	
	f. 157 (f)	"Noetic" roots
		referred to in MS
p. 110 (Pr.25)	-	$(-\sqrt{-1}, +\sqrt{-1})$
pp. 110–111 (Pr.26)	f. 201 (f)	
pp. 111–112 (Pr.27)	f. 160 (f), f. 162 (f)	
p. 112 (Pr.28)	f. 202 (f)	
pp. 112–113 (Pr.29)	-	
pp. 113–114 (Pr.30)	-	
pp. 114–115 (Pr.31)	f. 146 (f)	
p. 115 (Pr.32)	f. 146 (f)	
pp. 115–116 (Pr.33)	f. 148 (f)	
p. 116 (Pr.34)	-	
	(Many, many more	
	with no connection to	
	Praxis)	

Numerical Exegesis	**Add. MS 6782**
pp. 117–118 (Pr.1)	-
pp. 119–121 (Pr.2)	f. 398 (2)
p. 121–122 (Pr.2, devolution)	f. 399 (1)
p. 122–123 (Pr.2, devolution)	f. 397 (3)
pp. 124–125 (Pr.3)	(PRO), f. 13 (b. 1)
pp. 125–127 (Pr.3, anticipation)	(PRO), f. 12 (b. 2)
pp. 127–128 (Pr.3, rectification)	(PRO), f. 11 (b. 3)
pp. 128–130 (Pr.4)	f. 417 (c. 1)
	f. 416 (c. 2)
p. 130 (short-cut)	Add. MS 6782, f. 62

pp. 131–132 (Pr.5)   -
pp. 132–136 (Pr.6)   -
pp. 136–138 (Pr.7)   f. 396 (4)
pp.138–139 (Pr.7, devolution)   f. 394 (6)
pp. 141–142 (Pr.8)   (PRO), f. 9 (b. 4)
pp. 142–143 (Lemma)   -
pp. 143–145 (anticipation)   (PRO), f. 8 (b. 5)
pp. 145–146 (rectification)   (PRO), f. 7 (b. 6)
pp. 146–148 (Pr.9)   f. 415 (c. 3), f. 414 (c. 4)
                                        f. 413 (c. 5), f. 412 (c. 6)

pp. 149–150 (Pr.10)   -
p. 151 (Lemma)   -
pp. 151–152 (Pr.11)   -
pp. 153–1555 (Pr.12)   -
pp. 155–158 (Pr.13)   -
pp. 158–160 (Pr.14)   -
pp. 160–162 (Pr.15)   -
pp. 162–164 (Pr.16)   (PRO), f. 1 (b. 12)
pp. 164–167 (Approximations)   -
p. 168–180 Rules for guidance   -

Select Bibliography

Primary Sources

Aristotle, *Metaphysics*, Vol. XIII, tr. and ed. H. Tredennik,
[1977]
Harriot (Thomas). Unpublished MS. (British Library, Add. MS 6782, 6783,
6784). Also PRO Chichester.
Harriot (Thomas)
[1631] *Artis Analyticae Praxis*. Ed. W. Warner, London
Torporley (Nathaniel), *Papers*, Sion College. MS. Arc. L. 40.2/L.40. (Held at
Lambeth Palace Library, London.)
Viète (François)
[1646] *Opera Mathematica*, ed. Schooten, Leiden.
[1600] *De Numerosa potestatum resolutione*.
Wallis (J)
[1685] *A treatise of algebra, both historical and practical*. London.

Secondary Sources

Fauvel (John)
[1996] J.J. Sylvester and the papers of "Old Father Harriot", *The Harrioteer*.
Klein (J)
[1968] *Greek Mathematical Thought and the Origin of Algebra* (trans.
E. Brann), Cambridge, Massachusetts.
Knorr (Wilbur R)
[1986] *The Ancient Tradition of Geometric Problems*. Dover, New York.
Mahoney (MS)
[1994] *The Mathematical Career of Pierre de Fermat* (1601–1665), Princeton
U.P.

Pepper (Jon V)

[1967] "The Study of Thomas Harriot's Manuscripts," II. Harriot's Unpublished Papers, *History of Science, 6*, pp. 17–40.

Shirley (JW)

[1983] *Thomas Harriot: a biography* and Stedall, J.A. (2003)

[2000] "Rob'd of Glories – The Posthumous Misfortunes of Thomas Harriot and his Algebra", *Arch. Hist. Exact. Sci, 54*, pp. 455–497.

[2003] *The Greate Invention of Algebra. Thomas Harriot's Treatise on Equations*, Oxford University Press.

Tanner (Rosalind CH)

[1967a] "Thomas Harriot as Mathematician. A Legacy of Hearsay," *Physis* IX, pp. 235–256, 257–292.

[1967b] "The Study of Thomas Harriot's Manuscripts", I. Harriot's Will, *History of Science, 6*, pp. 1–16.

[1969] "Nathaniel Torporley and the Harriot Manuscripts," *Annals of Science, 25*, 339–349.

Witmer (T. Richard)

[1983] *The Analytic Art* (tr. of Viète), Kent State U.P.

Index to Introduction and Commentary

A

'A quadratum' 7–8
acephalic square 258
adventitious equation 244
affected power 262
algebra
    algebraic logic 6
    completely symbolic notation 9
    Harriot's role 1
    purely symbolic notation 1
    Renaissance to the modern world 15
    zero, introduction of 6
algorithmic methods 15
analytical geometry 211
'anticipation' 258
antithesis 215
Apollonius 212
application 215–16
Archimedes 212
Artis Analyticae Praxis
    achievements 11–14
    analysis exclusively algebraic 10–11
    Canonical equations generated from binomial roots 11, 12
    clear symbolism 13
    comparisons with MS and Torporley's copy 214
    contents compared with Harriot manuscript pages 287–91
    contents overview 11–14
    Definitions - author 10
    Definitions - purpose of providing 12
    equations solved, comparative table 263–69

errata v, 271–8
exponential notation lacking 15
first published v
first published algebraic work to be purely symbolic 15
focus on the structure of equations 15
generalization through lists of examples 215
Harriot's intentions v
Lemmas in section five 233–8
model polynomials 253
numerical exegesis 254–5
numerical solution of equations by successive approximation 13
ordering of sections 13
polynomial equations with numerical coefficients 11
Preface 10
purely symbolic notation 1
Rules for Guidance 13
section contents summarised 29
significance 14–16
structure of book 13
superior notation demonstrated 258, 259
text and surviving Harriot manuscripts 4
Torporley, Nathaniel 4
Warner, Walter as editor 4
Warner, Walter changes 286–7
avulsed powers 259, 261

B

backward composition 256, 257, 260
bibliography 293–4

biquadratic equations
    general method to find roots 13–14
    reduction 247
    removing the term of second highest
        degree 239
    table of derivation from 'originals'
        221–2
Bombelli, Raffaelo 7

C
Canonical equations
    background by Warner 22
    definitions 5n2, 217
    generated by binomial factors 11, 243
    primary reduced to secondary 223
    roots of Primary and Secondary 219
    secondary 254
    treatment 217
Canonical forms, pattern of formation 228
Canonical polynomial, definition 5n2
Cardano, Girolamo 6, 20, 234, 242
Cavendish, Charles 4
Chuquet, Nicolas 7
Clavious, Christopher 13
comma used as bracket 214
common equations 223
complex roots of equations 219–21, 235
conceiving and imagining 9
conjugate complex numbers 6, 246–7
conjugate equations, pairs of 247, 248
Copernicanism 2
corrective device 258–9
Cossic notation 13, 15, 213
Cossic numbers 6
cubic equations
    derivation from 'originals' 222
    fully solved 239, 242
    general method to find roots 13–14
    negative coefficients 240
    pure cube 260
    reciprocal 222
    removing the term of second highest
        degree 14, 239
    symbolic solution 245

D
De Radicalibus 215, 287
Dee, John 2

Definitions in *Artis Analyticae Praxis*
    1. specious logistic 23, 209–11
    2. equation 23
    3. synthesis 23–4, 211
    4. analysis 24, 211–12
    5. composition, resolution 24
    6. Zetetic [analysis] 24–5, 212
    7. Exegetic [analysis] 25, 212
    8. Poristic [analysis] 25, 212
    9. specious Exegesis 25–6
    10. numerical Exegesis 26
    11. secondary Exegesis 26
    12. root, value 26–7
    13. common or adventitious equation
        27
    14. originals of canonical equations 27
    15. primary canonical equations 27–8
    16. secondary canonical equations 28
    17. canonical equations 28–9
    18. reciprocal equation 29
degree of an equation, equality with number
        of roots 15, 219
Descartes, René
    algebraic geometry 10
    geometrism 15–16
    La Gèomètrie 2, 7, 9
    Pappus' locus problem 210
    Rule of Signs 15, 235, 236
    unity 9–10
Digges, Leonard 2
Digges, Thomas 2
Diophantus 6–7, 8, 20
discriminant of the cubic 235
dividing through an equation by known
        quantity 216
dividing through an equation by the un-
        known 216

E
epanorthosis 261
equations with numerical coefficients 254
equipollence 233, 235, 236, 287
exegesis [use of word] 253
explicated root of equation 229

F
'four-root law' 249
fractions reduced to lowest terms 214

G
Girard Albert 7
Grenville, Sir Richard 3

H
Hakluyt, Richard 2
Halley, Edmund 14, 237
Harriot, Thomas
 accomplishments 2
 accomplishments set out by
  D. T. Whiteside 2
 algebra transformed 1
 analytical geometry 211
 background 2
 background by Warner 19–20, 22
 career 3
 comparison with Viète 8–11
 completely symbolic notation 9
 conceptual connection with Viète 10
 equating all the terms of an equation
  to zero 13
 equations with numerical coefficients
  254
 exegesis [use of word] 253
 exponential notation 15
 facility in symbolic thinking 224
 influence on later English mathematicians 4
 manuscript papers 254
 notation revolutionary 1
 polynomial equations generated by
  product of binomial factors 1
 polynomial equations with terms
  equated to zero 1
 purely symbolic notation 1, 15
 reputation 1
 superior notation demonstrated 258,
  259
 telescopes 2
 two chief discoveries 5
 use of Viète's examples 253
 Will 3
Harriot, Thomas - papers
 Add. MS 6782 contents list 279–1
 Add. MS 6783 contents list 281–2
 Add. MS 6784 contents list 282
 comments on numbering 286–7
 contents and comparison with
  Torporley 283–5

 contents and numbering 285–6
 disposal after death 3–4, 3n1
 Torporley access 11
 'waste' 3
headless quadratic method 258
homogeneous term 13, 217, 247, 248
homogeneity, laws of 260
homogeneity, problems of 9
homogenous form of equations 213
Hues, Robert 3
hypobibasm 215, 216

I
imaginary roots of equations 12, 219–20,
  249
imagining and conceiving 9

K
Kepler, J 2

L
Lagrange, J L 6, 14, 235
Lower, William 6

M
Macraelius 217
magnitudes
 algebra of 9
 Descartes 10
 of dimension 10
 Viète, François 8, 209–10
mercantile capitalism 2

N
negative roots changed to positive roots 247,
  249
negative roots of equations 6, 12, 219–20,
  229
 existence recognised but usefulness
  challenged 222
noetic 221
Northumberland, 9th Earl of 3
notation
 completely symbolic notation 9
 Cossic 6, 13, 15, 213
 cube root 245
 Diophantine 6–7
 division line 223
 dots v, 215

equal signs 216
equality signs 214, 229
exponents 7, 15
   Harriot's unique 15
inequality sign v, 1
inequality signs 214, 216
letters for positive numbers 213
literal 4
literal sign for a general number 6
logic embodied in notation 1
multiplication sign 13, 213, 215, 218
negative sign 7, 258
purely symbolic 1, 15
revolutionary 1
separate signs for unknown and each
   power 6–7
square number 7–8
superscripts for collecting like terms
   224, 227
Viète, François 4–5, 7, 8
Numerical Exegesis 21, 22
numerical logistic, Viète's definition 8–9,
   209

O
operation rules, François Viète 209–10
'originals'
   biquadratic equations derived from
      them 221–2
   cubic equations derived from them
      218, 222
   treatment 217, 218

P
Pacioli, Luca 6
Pappus 8
Pappus' locus problem 210
parabolismus 215, 216
Pell, John 4
Pepper, Jon V 1
Plato 211
polynomial equations with numerical
   coefficients
   solution by successive approximations
      11
   use by Harriot of Viète's work 9
poristic verification 259, 260, 261
posited root of equation 229

positive roots changed to negative roots 247,
   249
positive roots of equations 219
privative roots of equations 229
problems (section 3) related to corresponding propositions (section 4) table including roots of equations
   230–2
problems vs propositions 217
Proclus 215–16
propositions (section 4) related to corresponding problems (section 3)
   table including roots of equations 230–2
propositions vs problems 217
pure powers 256, 257

Q
quadratum acephalum method 258

R
radix 229
Ralegh (Raleigh), Walter 3
real roots from conjugate complex numbers
   246–7
rectification 261

S
specious arithmetic 21
specious logistic 8–9, 10
square root of negative quantity 245
Stedall, Jacqueline A 4
Stevin, Simon 7, 21
Stifel, Michael 6
successive approximations method 255–7
Sylvester, J J 6
symbolic arithmetic 21
symbolic logistic, Viète's definition 9, 209
symmetric functions of roots of an equation
   219, 260, 261

T
Tanner, R C H 1, 4, 14
Tartaglia, Niccolò 20
Theon 211, 212
Torporley, Nathaniel 3
Torporley, Nathaniel
   access to Harriot papers 11
   contents and comparison with Harriot
      papers 283–5

transposition under opposite signs 216
trial divisor 255, 257

U
unaffected equation 256
unity
    conceived as a number 7
    dimension of 9–10
    problem of homogeneity 9

V
verification by substitution 257
Viète, François
    analysis identified with algebra as well
        as geometry 8
    comparison with Harriot 8–11
    generality of algebra 8
    literal sign for a general number 6
    magnitudes 8
    notation 4–5, 7, 8

W
Wallis, John 4, 9, 233, 236, 238
Warner, Walter
    changes made to Harriot material
        287–8

conceptual connection with Viète 10
connection with Harriot 3
debt due to him 16
editor of *Artis Analyticae Praxis* 4,
    10–12
eulogy to Harriot 11
Harriot's two chief discoveries 5
order of sections of *Artis Analyticae
    Praxis* 13
Whiteside, D T 1, 2
Witmer, T Richard 9

Z
zero
    equating all the terms of an equation
        to zero 13, 218–19
    introduced into algebra 6
    use as a calculable quantity 241
    use by Harriot 213

Sources and Studies in the
History of Mathematics and Physical Sciences

Continued from page ii

C.C. Heyde/E. Seneta
I.J. Bienaymé: Statistical Theory Anticipated

J.P. Hogendijk
Ibn Al-Haytham's *Completion of the Conics*

J. Høyrup
Length, Widths, Surfaces: A Portrait of Old Babylonian Algebra and Its Kin

A. Jones (Ed.)
Pappus of Alexandria, Book 7 of the *Collection*

E. Kheirandish
The Arabic Version of Euclid's *Optics,* Volumes I and II

J. Lützen
Joseph Liouville 1809–1882: Master of Pure and Applied Mathematics

J. Lützen
The Prehistory of the Theory of Distributions

G.H. Moore
Zermelo's Axiom of Choice

O. Neugebauer
A History of Ancient Mathematical Astronomy

O. Neugebauer
Astronomical Cuneiform Texts

F.J. Ragep
Nasr al-Dn al-Ts's *Memoir on Astronomy*
(al-Tadhkira fi cilm al-hay'a)

B.A. Rosenfeld
A History of Non-Euclidean Astronomy

G. Schubring
**Conflicts between Generalization, Rigor, and Intuition: Number Concepts
Underlying the Development of Analysis in 17-19th Century France and Germany**

M. Seltman/R. Goulding
**Thomas Harriot's Artis Analyticae Praxis: An English Translation with
Commentary**

Continued from previous page

J. Sesiano
Books IV to VII of Diophantus' *Arithemetica***: In the Arabic Translation Attributed to Qustā ibn Lūqā**

L. Sigler
Fibonacci's Liber Abaci: A Translation into Modern English of Leonardo Pisano's Book of Calculation

B. Stephenson
Kepler's Physical Astronomy

N.M. Swerdlow/O. Neugebauer
Mathematical Astronomy In Copernicus' De Revolutionibus

G.J. Toomer (Ed.)
Apolonius *Conics* **Books V to VII: The Arabic Translation of the Lost Greek Original in the Version of Banū Mūsă**, Edited, with English Translation and Commentary by G.J. Toomer

G.J. Toomer (Ed.)
Diocles on Burning Mirrors: The Arabic Translation of the Lost Greek Original, Edited, with English Translation and Commentary by G.J. Toomer

C. Truesdell
The Tragicomical History of Thermodynamics, 1822–1854

I. Tweddle
James Stirling's Methodus Differentialis: An Annotated Translation of Stirling's Text

K. von Meyenn/A. Hermann/V.F. Weiskopf (Eds.)
Wolfgang Pauli: Scientific Correspondence II: 1930–1939

K. von Meyenn (Ed.)
Wolfgang Pauli: Scientific Correspondence III: 1940–1949

K. von Meyenn (Ed.)
Wolfgang Pauli: Scientific Correspondence IV, Part I: 1950–1952

K. von Meyenn (Ed.)
Wolfgang Pauli: Scientific Correspondence IV, Part 2: 1953–1954

J. Stedall
The Arithmetic of Infinitesimals: John Wallis 1656